I0038708

Steel Metallurgy

Steel Metallurgy

Molly Hamilton

Larsen & Keller
www.larsen-keller.com

Steel Metallurgy
Molly Hamilton
ISBN: 978-1-64172-678-8 (Hardback)

© 2022 Larsen & Keller

▤ Larsen & Keller

Published by Larsen and Keller Education,
5 Penn Plaza,
19th Floor,
New York, NY 10001, USA

Cataloging-in-Publication Data

Steel metallurgy / Molly Hamilton.
        p. cm.
Includes bibliographical references and index.
ISBN 978-1-64172-678-8
1. Steel--Metallurgy. 2. Metallurgy. 3. Steel. I. Hamilton, Molly.
TN705 .S84 2022
669.142--dc23

This book contains information obtained from authentic and highly regarded sources. All chapters are published with permission under the Creative Commons Attribution Share Alike License or equivalent. A wide variety of references are listed. Permissions and sources are indicated; for detailed attributions, please refer to the permissions page. Reasonable efforts have been made to publish reliable data and information, but the authors, editors and publisher cannot assume any responsibility for the validity of all materials or the consequences of their use.

Trademark Notice: All trademarks used herein are the property of their respective owners. The use of any trademark in this text does not vest in the author or publisher any trademark ownership rights in such trademarks, nor does the use of such trademarks imply any affiliation with or endorsement of this book by such owners.

For more information regarding Larsen and Keller Education and its products, please visit the publisher's website www.larsen-keller.com

Table of Contents

Permissions

Index

Preface

The field of materials science and engineering which studies the physical and chemical behavior of metallic elements is called metallurgy. It also studies their inter-metallic compounds and their mixtures, which are known as alloys. Steel metallurgy is a domain under the subfield of metallurgy known as ferrous metallurgy. Steel is an alloy of iron and carbon in which the carbon content is not more than 2 percent. There are many types of steel which are classified broadly into a few major groups on the basis of specific criteria. These are surface conditions, chemical compositions, applications and shapes. This book provides comprehensive insights into the field of steel metallurgy. The fundamentals as well as modern approaches of this field are discussed in it. Those with an interest in the field of steel metallurgy would find this book helpful.

Given below is the chapter wise description of the book:

Chapter 1- Iron is a metallic chemical element. It can be used to form a number of different alloys and compounds such as fernico, cast iron, pig iron, cementite and austenite. This chapter has been carefully written to provide an easy understanding of the varied facets of iron as well as its alloys.

Chapter 2- Steel refers to an alloy of iron with carbon and other elements. There are numerous types of steel such as carbon steel, alloy steel, tool steel, stainless steel and structural steel. Carbon steel has three major sub-types, namely, low carbon, medium carbon and high carbon steel. This chapter discusses in detail these types of steel as well as the processes involved in the treatment of steel.

Chapter 3- The physical properties of steel include high strength, resistance to corrosion and low weight. The mechanical properties of steel are its weldability, durability, ductility and malleability. The topics elaborated in this chapter will help in gaining a better perspective about these mechanical and physical properties of steel.

Chapter 4- The process of producing steel from iron ore or scrap is called steelmaking. It is broadly divided into two steps, namely, primary steelmaking and secondary steelmaking. This chapter has been carefully written to provide an easy understanding of these facets of the steelmaking process.

Chapter 5- Steel is a versatile material which is used in diverse areas. Some of them are construction, automotive and transport sectors, packaging food and catering, and energy production. The chapter closely examines the applications of steel in these areas to provide an extensive understanding of the subject.

At the end, I would like to thank all those who dedicated their time and efforts for the successful completion of this book. I also wish to convey my gratitude towards my friends and family who supported me at every step.

Molly Hamilton

Iron and its Alloys

Iron is a metallic chemical element. It can be used to form a number of different alloys and compounds such as fernico, cast iron, pig iron, cementite and austenite. This chapter has been carefully written to provide an easy understanding of the varied facets of iron as well as its alloys.

Iron

Metallic iron was not obtained from its naturally occurring compounds at so early a date as some of the other metals, especially copper and tin. This is due to its high point of fusion, and to the much greater difficulty in obtaining it in the metallic state from its compounds. Thus, in prehistoric times iron does not appear till after bronze, i.e. mixtures containing copper as essential constituent, and was apparently at first a great rarity.

Notwithstanding the wide distribution of iron, it scarcely ever occurs in the metallic state on account of its tendency to form compounds with oxygen and sulphur. The chief occurrence of metallic iron, except in some rather accidental cases through the action of chemical processes connected with volcanic activity, is in certain meteorites. These are masses which do not originally belong to the earth, but which, in the course of their flight through space, approach so closely to the earth that, owing to atmospheric friction, they lose their kinetic energy, which is thereby converted into heat, and fall to the earth. Many of these masses consist of iron.

Masses of native iron also occur, although rarely (e.g. at Of vivak in Greenland), whose meteoric origin is doubtful, although no explanation has been given of any other possible origin.

Iron is a grey, tenacious metal, which fuses with great difficulty, at about 1600°; it combines with free oxygen quickly at high temperatures, slowly at low ones. In the heat essentially compounds of the formula Fe_3O_4 to Fe_2O_3 are formed; in the cold, iron hydroxide, $Fe(OH)_3$, is formed. The hydrogen necessary for this is taken up in the form of water; in fact, iron "rusts" or oxidises at a low temperature only in moist, not, or not measurably, in dry air. Since the rust does not cohere, it does not protect the iron against further oxidation.

At all temperatures water is decomposed by iron. The decomposition of water by red-hot iron is a classical experiment. Even at the ordinary temperature decomposition takes place with evolution of hydrogen, but exceedingly slowly, so that the evolution of hydrogen can be observed only by using large surfaces (iron powder). Iron is dissolved even by the weakest acids, thereby passing into divalent diferrion with evolution of hydrogen.

Commercial Iron

Commercial iron is not pure, but contains up to as much as 5 per cent of carbon, which has a very great influence on its properties, and also smaller quantities of other impurities. While pure iron, although very tenacious, is comparatively soft, its hardness increases with the amount of carbon it contains, and its behaviour at moderately high temperatures becomes essentially different.

There are three chief kinds of commercial iron, viz. wrought-iron, steel, and cast-iron; the first contains the smallest, the last the highest, amount of carbon. Wrought-iron approximates most nearly both in composition and in properties to pure iron; it is tough, not very hard, and on being heated first becomes soft like wax or sodium before melting. This property is of the greatest importance for the technical working of iron, as it renders it possible to shape the metal and to unite different pieces without it being necessary to raise the temperature to the melting point of the metal. On the contrary, it is sufficient to heat to the temperature of softening (about 600°), so as to attain the object by pressing, rolling, and forging. The uniting of the two pieces of iron by pressure (hammering) is called welding. The temperature necessary for this is bright red-heat.

The properties of wrought-iron do not undergo essential change when it is heated and suddenly cooled. The character of steel, however, depends in the highest degree on such treatment.

Steel is iron which contains from 0.8 to 2.5 per cent of carbon, but is otherwise as pure as possible. The carbon is chemically combined with the iron, and this carburetted iron or iron carbide, Fe_3C, is alloyed with the rest of the iron. The result of the presence of this foreign substance is, in the first place, an appreciable sinking of the melting point; at 1400° steel is liquid and can be cast. Cast-steel is a metal consisting of fine crystalline grains, which, like wrought-iron, softens before melting, and can therefore be forged. By such treatment steel acquires a fibrous or sinewy character, similar to wrought-iron. If the steel is made red hot and then suddenly cooled, it becomes brittle, and at the same time acquires its highest degree of hardness. It is then so hard that it scratches glass, and is hence called glass hard. If this steel is again carefully heated, all degrees of hardness can be imparted to it, for it increases in softness the longer or the higher it is heated. This process is called the tempering of steel.

As an index of the degree of tempering to be attained, use has been made from olden times of the colours which a bright steel surface acquires on being heated. At about 220°, the metal begins to oxidise in the air with a measurable velocity, and the oxide produced forms a thin coating on the metal. If the thickness of this coating is of the order of a wave-length of light, the corresponding interference colours, or the " colours of thin plates," begin to appear. Since the shortest of the visible waves, the violet, is first extinguished, the first tarnish- colour to appear is the complementary colour, pale straw-yellow. This passes through the colours orange, purple, violet, blue, and finally becomes grey. To each of these colours there corresponds a definite degree of hardness of the steel. Steel for tools to work iron is allowed to reach the yellow stage, for brass the purple-red stage, while tools for wood are allowed to become blue. Although colour and hardness do not exactly correspond, still the correspondence is sufficient for an experienced workman.

The great utility of steel in the arts is due to the diversity in the degrees of hardness which it can acquire. In the soft state it can be shaped to any desired form, and the shaped objects can then be brought to any degree of hardness.

It is only in recent years that the theory of tempering has been made clear. Iron carbide, Fe_3C, is not only itself very hard, but it forms with pure iron a homogeneous mixture, a " solid solution," which is also hard; so much the less hard, the less carbide it contains. If, now, such a solid solution, consisting at higher temperatures of carbide and iron, is slowly cooled, it breaks up at about 670° into pure iron and iron carbide, which exist as a conglomerate side by side. Since pure iron is soft, it imparts this property also to the mixture.

If, however, the cooling is performed rapidly, the breaking up of the solid solution does not occur, and the latter therefore preserves its hardness. The solid solution hereby becomes metastable or to a certain extent supersaturated.

This explains, in the first place, why quenched steel is hard, while slowly cooled steel is soft. The tempering of hard steel, now, consists in the separation of the solid solution into its two constituents through elevation of the temperature, the separation occurring all the more rapidly the higher the temperature. By sudden cooling, the state of the mixture attained at any point is preserved, since, at the ordinary temperature, the velocity of change is immeasurably small. The corresponding degree of hardness is then obtained.

These considerations also make clear the fact, learned by experience, that the temper depends not only on the temperature but also on the time, in such a way that a lower temperature for a long period has the same effect as a higher temperature for a shorter time.

The tempering can be carried out in one operation by appropriately heating to above 670° until the desired mixture of iron and solid solution (the equilibrium between which alters with the temperature) is produced, and then fixing this state by suddenly cooling. The temperature necessary for obtaining a definite degree of hardness depends on the amount of carbon present. If this is known, the temperature required to produce a given degree of hardness can be decided beforehand.

If the amount of carbon increases to from 4 to 5 per cent, the melting point of the iron becomes still lower, and the metal loses its toughness and the power of assuming the fibrous condition, but it still retains the power of being tempered to a certain degree. Such iron is called castiron.

Two kinds of cast-iron are distinguished, white and grey. The former is obtained by quickly cooling; it is very hard and crystalline, and contains the greater part of its carbon chemically combined as carbide. When the cast-iron is slowly cooled, part of the carbon separates out in fine laminae as graphite, which imparts a grey colour to the iron. At the same time the metal becomes less hard and brittle, and the grain finer. In this condition cast-iron is used for innumerable purposes where ease in the shaping of the object by casting has to be taken into account, and where the smaller resistance of the metal to pulling strain and bending is no essential drawback.

Iron Occurrence

Meteorite iron is not pure. Usually it contains 30% or less other elements. It may be hammered in the cold state, but becomes brittle when heated.

Iron inclusions may be found in basalt and other igneous rocks. It is supposed that find the Earth's core is predominantly iron metal starting from 2900 km depth (the Earth's equatorial radius is 6377 km) and is composed of iron (91-92%)/nickel (8-9%) alloy.

Iron is most abundant element after oxygen, silicon and aluminium. Iron compounds form a great number of minerals and rocks, soils and living organisms; however the iron ores are few. Iron oxide ores are most valuable among them.

Iron is essential to nearly all known organisms, with concentration approximately 0.02%. It is a very essential element for oxygen exchange and oxidizing processes. Some living organisms, so-called iron concentrators are able to deposit large amounts of iron (until 17-20%) within them like, for instance, iron depositing bacteria. Iron in living organisms is almost entirely involved in protein processes. Iron deficiency, aggravated by high soil pH (alkali soils), brings to plants growth inhibition and chlorosis, a condition in which leaves produce insufficient chlorophyll. Iron abundance in the case of low soil pH (acid soils) is also harmful: it causes flowers sterility. Such plant diseases may occur in large areas.

Iron Alloys

Iron alloy is an alloy that have iron as the principal component. Iron is used as a constituent in most of the commercial alloys. For example, iron is the major component of wrought and cast iron and wrought and cast steel. Iron can be alloyed with manganese, silicon, vanadium, chromium, molybdenum, niobium, selenium, titanium, phosphorus, or other elements for commercial use. Some iron alloys are also used as addition agents in steel-making. There are special-purpose iron alloys that demonstrate exceptional characteristics, like magnetic properties, electrical resistance, thermal expansion, corrosion resistance, and heat resistance, etc.

Steel Alloy is considered to be the most popular amongst the Iron Alloys and widely used across the industries.

The Production of Iron Alloys

Iron alloy components have been produced from the liquid state for many centuries with mass production of steel commencing in the late nineteenth century. Molten steel was traditionally cast into standing molds but continuous casting was developed in the 1950s which enabled the production of huge tonnages of steel. Most of the world's steel production is now obtained through this casting route. Following casting, as-cast slabs are thermal and mechanical processed to the desired final shape. In the last few decades, however, near-net-shape casting processes have been developed that are capable of producing near-final-shape products directly from liquid with secondary processing such as hot rolling reduced to an absolute minimum. On a reduced scale, there are various types of novel iron alloys produced in the amorphous, nanocrystalline, and intermetallic form. These materials are either manufactured by conventional casting or by rapid solidification processing, spray deposition, mechanical attrition, and sintering or physical and chemical vapor deposition, etc. Iron alloys produced by these more exotic techniques are generally restricted to specific applications.

Iron Alloy Castings

Iron castings are produced by molding iron alloys or molten iron. Gray iron is possibly one of the oldest worked metal used for casting. The metal is also one of the most abundant and least expensive material.

Gray iron was the original cast iron, though it has recently been replaced in various applications by other iron-carbon alloys with higher tensile strength. Ductile iron, as the name suggests, has high ductility than traditional iron materials, such as gray iron. These materials incline to be brittle and are prone to fracture under high tensile stress.

Iron is cast as like any other metal. It is poured into a mold and extracted after cools down. There are different types of casting methods, through which iron can be molded, which include shell molding, green sand molding, and centrifugal molding. The procedure through which the iron is caste can effect the mechanical properties of the metal, especially in regard to its cooling rate. Iron castings are used in a variety of applications, including - automotive, appliance, agricultural, and machinery industries. Cast iron components are used pump housings, engine blocks, electrical boxes, decorative castings, and more.

Fernico

Fernico describe a family of metal alloys made primarily of iron, nickel and cobalt. The family includes Kovar, FerNiCo I, FerNiCo II, and Dumet. The name is made up of the chemical symbols of its constituent three elements. "Dumet" is a portmanteau of "dual" and "metal," because it is a heterogeneous alloy, usually fabricated in the form of a wire with an alloy core and a copper cladding. These alloys possess the properties of electrical conductivity, minimal oxidation and formation of porous surfaces at working temperatures of glass and thermal coefficients of expansion which match glass closely. These requirements allow the alloys to be used in glass seals, such that the seal does not crack, fracture or leak with changes in temperature. Dumet is most commonly used in seals where lead-in wires pass through the glass bulb wall of standard household electric lamps (light bulbs) among other things. The two Fernico alloys both consist of iron (Fe), nickel (Ni), and cobalt (Co). Fernico is used at high temperatures (20 to 800°C) and is identical to Kovar. Fernico II is used at cryogenic temperatures in the -80. -180°C range. Both are used to create electrically conductive paths through the walls of sealed borosilicate glass containers. Dumet is used for a similar purpose, but is tailored for seals through soda lime and lead alkali silicate glasses. These alloys adhere to lead-tin, tin, and silver solders. Other metals, including copper, molybdenum, nickel, and steel can be spot-welded to the FerNiCo alloys forming low resistance electrical connections.

Typical Compositions

Given in weight %

	Fe	Ni	Co	C	Si	Mn
FerNiCo I/Kovar	balance	29%	17%	<0.01%	0.2%	0.3%
FerNiCo II	balance	31%	15%	<0.01%	0.2%	0.3%
Dumet Core	58%	42%	-	-	-	-

FerNiCo I has the same linear coefficient of expansion as certain types of borosilicate ("hard" glass), (c6.5 × $10^{-6 \circ}K^{-1}$, thus serving as an ideal material for the lead-out wires or other seal structures in light bulbs and thermionic valves. Dumet is also used for this purpose, but for passing through softer soda-lime and lead-alkali glasses. This wire is often coated with a glass-like film of sodium metaborate, ($NaBO_2$), so the molten glass will "wet" and adhere to it. 25% by mass of the finished wire is copper.Cunife exhibits a similar property.

Uses

There are very few uses of Fernico. Some of them are:

- It is used to seal metals and glass.
- It is often used in the form of nanopowder.

Cast Iron

Cast iron is an iron-carbon alloy. It contains 2 to 4 percent carbon, along with varying amounts of silicon and manganese and traces of impurities such as sulfur and phosphorus. It is made by reducing iron ore in a blast furnace. The liquid iron is cast, or poured and hardened, into crude ingots called pigs, and the pigs are subsequently remelted along with scrap and alloying elements in cupola furnaces and recast into molds for producing a variety of products.

The Chinese produced cast iron as early as the 6th century BC, and it was produced sporadically in Europe by the 14th century. It was introduced into England about 1500; the first ironworks in America were established on the James River, Virginia, in 1619. During the 18th and 19th centuries, cast iron was a cheaper engineering material than wrought iron because it did not require intensive refining and working with hammers, but it was more brittle and inferior in tensile strength. Nevertheless, its load-bearing strength made it the first important structural metal, and it was used in some of the earliest skyscrapers. In the 20th century, steel replaced cast iron in construction, but cast iron continues to have many industrial applications.

Most cast iron is either so-called gray iron or white iron, the colours shown by fracture. Gray iron contains more silicon and is less hard and more machinable than is white iron. Both are brittle, but a malleable cast iron produced by a prolonged heat treatment was developed in France in the

18th century, and a cast iron that is ductile as cast was invented in the United States and Britain in 1948. Such ductile irons now constitute a major family of metals that are widely used for gears, dies, automobile crankshafts, and many other machine parts.

Mechanical Properties of Cast Iron

The mechanical properties of a material indicate how it responds under specific stresses, which helps to determine its suitability for different applications. Specifications are set by organizations such as the American Society for Testing and Materials (ASTM) so that users can purchase materials with confidence that they meet the requirements for their application. The most commonly used cast gray iron specification is ASTM A48.

In order to qualify cast products according to their specifications, a standard practice is to cast a test bar along with the engineered castings. The ASTM tests are then applied to this test bar and the results are used to qualify the entire batch of castings.

Specifications are also important when welding cast iron parts together. The weld must meet or exceed the mechanical properties of the material being welded together—otherwise, fractures and failures can occur.

When welding, it's critical that the weld meets or exceeds the mechanical properties of the material to prevent fractures and failures.

A few common mechanical properties for cast iron include:

- Hardness – material's resistance to abrasion and indentation.

- Toughness – material's ability to absorb energy.

- Ductility – material's ability to deform without fracture.

- Elasticity – material's ability to return to its original dimensions after it has been deformed.

- Malleability – material's ability to deform under compression without rupturing.

- Tensile strength – the greatest longitudinal stress a material can bear without tearing apart.

- Fatigue strength – the highest stress that a material can withstand for a given number of cycles without breaking.

This table summarizes some of the key mechanical properties for various grades of cast iron.

	Brinell Hardness	Tensile Strength	Modulus of Elasticity	% Elongation (In 50 Mm)
Gray iron class 25	187	29.9 ksi	16.1 Msi	–
Gray iron class 40	235	41.9 ksi	18.2 Msi	–
Ductile iron grade 60-40-18	130 – 170	60 ksi	24.5 Msi	–
Ductile iron grade 129-90-02	240 – 300	120 ksi	25.5 Msi	–
CGI grade 250	179 max	36.2 ksi min		3
CGI grade 450	207 – 269	65.2 ksi min		1

Common Applications of Cast Iron

The various properties of different types of cast iron results in each type being suited for specific applications.

Gray Iron Applications

One of the key characteristics of gray iron is its ability to resist wear even when lubrication supply is limited (e.g. the upper cylinder walls in engine blocks). Gray iron is used to make engine blocks and cylinder heads, manifolds, gas burners, gear blanks, enclosures, and housings.

White Iron Applications

The chilling process used to make white iron results in a brittle material that is very resistant to wear and abrasions. For this reason, it is used to make mill linings, shot-blasting nozzles, railroad brake shoes, slurry pump housings, rolling mill rolls, and crushers.

Ni-Hard Iron is specifically used for mixer paddles, augers and dies, liner plates for ball mills, coal chutes, and wire guides for drawing wires.

Abrasion-resistant white iron is used to produce a variety of machine parts such as slurry pump housings.

Ductile Iron Applications

Ductile iron itself can be broken down into different grades, each with their own property specifi-cations and most suitable applications. It is easy to machine, has good fatigue and yield strength, while being wear resistant. Its most well-known feature, however, is ductility. Ductile iron can be used to make steering knuckles, plow shares, crankshafts, heavy duty gears, automotive and truck suspension components, hydraulic components, and automobile door hinges.

Malleable Iron Applications

Different grades of malleable iron correspond to different microcrystalline structures. Specific attributes that make malleable iron attractive are its ability to retain and store lubricants, the non-abrasive wear particles, and the porous surface which traps other abrasive debris. Malleable iron is used for heavy duty bearing surfaces, chains, sprockets, connecting rods, drive train and axle components, railroad rolling stock, and farm and construction machinery.

Malleable iron is used for heavy-duty bearing surfaces such as drive
train and axle components.

Compacted Graphite Iron Applications

Compacted graphite iron is beginning to make its presence known in commercial applications. The combination of the properties of gray iron and white iron create a high strength and high thermal conductivity product—suitable for diesel engine blocks and frames, cylinder liners, brake discs for trains, exhaust manifolds, and gear plates in high pressure pumps.

Machining and Finishing

The hardness properties of cast iron demand careful selection of machine tool materials. Coated carbides are effective in production machining environments, but newer materials are being devel-oped continuously as technology improves.

Surface finishing of cast iron products varies greatly according to the use. A few common applications:

- Electroplating

- Hot-dipping

- Thermal spraying

- Diffusion coating

- Conversion coating

- Porcelain enamelling

- Liquid organic coating

- Dry powder organic coating

Alloying Elements

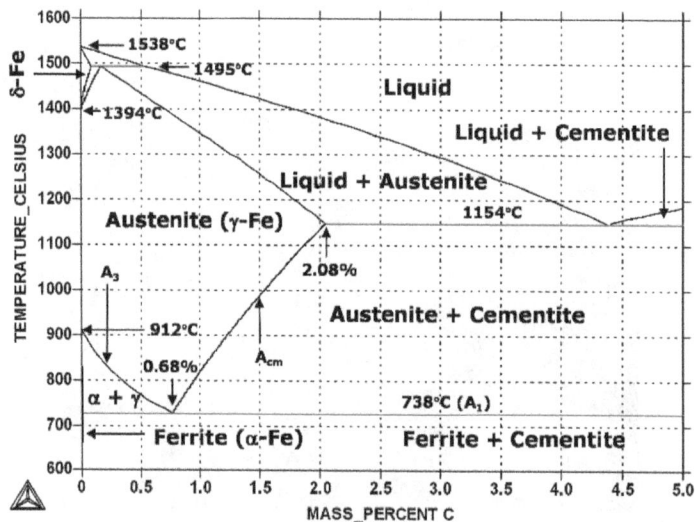

Iron-cementite meta-stable diagram.

Cast iron's properties are changed by adding various alloying elements, or alloyants. Next to carbon, silicon is the most important alloyant because it forces carbon out of solution. A low percentage of silicon allows carbon to remain in solution forming iron carbide and the production of white cast iron. A high percentage of silicon forces carbon out of solution forming graphite and the production of grey cast iron. Other alloying agents, manganese, chromium, molybdenum, titanium and vanadium counteracts silicon, promotes the retention of carbon, and the formation of those carbides. Nickel and copper increase strength, and machinability, but do not change the amount of graphite formed. The carbon in the form of graphite results in a softer iron, reduces shrinkage, lowers strength, and decreases density. Sulfur, largely a contaminant when present, forms iron sulfide, which prevents the formation of graphite and increases hardness. The problem with sulfur is that it makes molten cast iron viscous, which causes defects. To counter the effects of sulfur, manganese is added because the two form into manganese sulfide instead of iron sulfide. The manganese sulfide is lighter than the melt, so it tends to float out of the melt and into the slag. The amount of manganese required to neutralize sulfur is $1.7 \times$ sulfur content + 0.3%. If more than this amount of manganese is added, then manganese carbide forms, which increases hardness and chilling, except in grey iron, where up to 1% of manganese increases strength and density.

Nickel is one of the most common alloying elements because it refines the pearlite and graphite structure, improves toughness, and evens out hardness differences between section thicknesses. Chromium is added in small amounts to reduce free graphite, produce chill, and because it is a powerful carbide stabilizer; nickel is often added in conjunction. A small amount of tin can be added as a substitute for 0.5% chromium. Copper is added in the ladle or in the furnace, on the order of 0.5–2.5%, to decrease chill, refine graphite, and increase fluidity. Molybdenum is added on the order of 0.3–1% to increase chill and refine the graphite and pearlite structure; it is often added in conjunction with nickel, copper, and chromium to form high strength irons. Titanium is added as a degasser and deoxidizer, but it also increases fluidity. 0.15–0.5% vanadium is added to cast iron to stabilize cementite, increase hardness, and increase resistance to wear and heat. 0.1–0.3% zirconium helps to form graphite, deoxidize, and increase fluidity.

In malleable iron melts, bismuth is added, on the scale of 0.002–0.01%, to increase how much silicon can be added. In white iron, boron is added to aid in the production of malleable iron; it also reduces the coarsening effect of bismuth.

Grey Cast Iron

Pair of English firedogs, 1576. These, with firebacks, were common early uses of cast iron, as little strength in the metal was needed.

Grey cast iron is characterised by its graphitic microstructure, which causes fractures of the material to have a grey appearance. It is the most commonly used cast iron and the most widely used cast material based on weight. Most cast irons have a chemical composition of 2.5–4.0% carbon, 1–3% silicon, and the remainder iron. Grey cast iron has less tensile strength and shock resistance than steel, but its compressive strength is comparable to low- and medium-carbon steel. These mechanical properties are controlled by the size and shape of the graphite flakes present in the microstructure and can be characterised according to the guidelines given by the ASTM.

White Cast Iron

White cast iron displays white fractured surfaces due to the presence of an iron carbide precipitate called cementite. With a lower silicon content (graphitizing agent) and faster cooling rate, the carbon in white cast iron precipitates out of the melt as the metastable phase cementite, Fe_3C, rather than graphite. The cementite which precipitates from the melt forms as relatively large particles. As the iron carbide precipitates out, it withdraws carbon from the original melt, moving the mixture toward one that is closer to eutectic, and the remaining phase is the lower iron-carbon austenite (which on cooling might transform to martensite). These eutectic carbides are much too large to provide the benefit of what is called precipitation hardening (as in some steels, where much smaller cementite precipitates might inhibit plastic deformation by impeding the movement of dislocations through the pure iron ferrite matrix). Rather, they increase the bulk hardness of the cast iron simply by virtue of their own very high hardness and their substantial volume fraction, such that the bulk hardness can be approximated by a rule of mixtures. In any case, they offer hardness at the expense of toughness. Since carbide makes up a large fraction of the material, white cast iron could reasonably be classified as a cermet. White iron is too brittle for use in many structural components, but with good hardness and abrasion resistance and relatively low cost, it finds use in such applications as the wear surfaces (impeller and volute) of slurry pumps, shell liners and lifter bars in ball mills and autogenous grinding mills, balls and rings in coal pulverisers, and the teeth of a backhoe's digging bucket (although cast medium-carbon martensitic steel is more common for this application).

It is difficult to cool thick castings fast enough to solidify the melt as white cast iron all the way through. However, rapid cooling can be used to solidify a shell of white cast iron, after which the remainder cools more slowly to form a core of grey cast iron. The resulting casting, called a *chilled casting*, has the benefits of a hard surface with a somewhat tougher interior.

High-chromium white iron alloys allow massive castings (for example, a 10-tonne impeller) to be sand cast, as the chromium reduces cooling rate required to produce carbides through the greater thicknesses of material. Chromium also produces carbides with impressive abrasion resistance. These high-chromium alloys attribute their superior hardness to the presence of chromium carbides. The main form of these carbides are the eutectic or primary M_7C_3 carbides, where "M" represents iron or chromium and can vary depending on the alloy's composition. The eutectic carbides form as bundles of hollow hexagonal rods and grow perpendicular to the hexagonal basal plane. The hardness of these carbides are within the range of 1500-1800HV.

Malleable Cast Iron

Malleable iron starts as a white iron casting that is then heat treated for a day or two at about 950 °C (1,740 °F) and then cooled over a day or two. As a result, the carbon in iron carbide transforms into graphite and ferrite plus carbon (austenite). The slow process allows the surface tension to form the graphite into spheroidal particles rather than flakes. Due to their lower aspect ratio, the spheroids are relatively short and far from one another, and have a lower cross section vis-a-vis a propagating crack or phonon. They also have blunt boundaries, as opposed to flakes, which alleviates the stress concentration problems found in grey cast iron. In general, the properties of malleable cast iron are more like those of mild steel. There is a limit to how large a part can be cast in malleable iron, as it is made from white cast iron.

Ductile Cast Iron

Developed in 1948, *nodular* or *ductile cast iron* has its graphite in the form of very tiny nodules with the graphite in the form of concentric layers forming the nodules. As a result, the properties of ductile cast iron are that of a spongy steel without the stress concentration effects that flakes of graphite would produce. Tiny amounts of 0.02 to 0.1% magnesium, and only 0.02 to 0.04% cerium added to these alloys slow the growth of graphite precipitates by bonding to the edges of the graphite planes. Along with careful control of other elements and timing, this allows the carbon to separate as spheroidal particles as the material solidifies. The properties are similar to malleable iron, but parts can be cast with larger sections.

Table of Comparative Qualities of Cast Irons.

Name	Nominal composition [% by weight]	Form and condition	Yield strength [ksi (0.2% offset)]	Tensile strength [ksi]	Elongation [% (in 2 inches)]	Hardness [Brinell scale]	Uses
Grey cast iron (ASTM A48)	C 3.4, Si 1.8, Mn 0.5	Cast	—	50	0.5	260	Engine cylinder blocks, flywheels, gearbox cases, machine-tool bases
White cast iron	C 3.4, Si 0.7, Mn 0.6	Cast (as cast)	—	25	0	450	Bearing surfaces
Malleable iron (ASTM A47)	C 2.5, Si 1.0, Mn 0.55	Cast (annealed)	33	52	12	130	Axle bearings, track wheels, automotive crankshafts
Ductile or nodular iron	C 3.4, P 0.1, Mn 0.4, Ni 1.0, Mg 0.06	Cast	53	70	18	170	Gears, camshafts, crankshafts
Ductile or nodular iron (ASTM A339)	—	cast (quench tempered)	108	135	5	310	—
Ni-hard type 2	C 2.7, Si 0.6, Mn 0.5, Ni 4.5, Cr 2.0	Sand-cast	—	55	—	550	High strength applications
Ni-resist type 2	C 3.0, Si 2.0, Mn 1.0, Ni 20.0, Cr 2.5	Cast	—	27	2	140	Resistance to heat and corrosion

Pig Iron

Pig iron is the product of smelting iron ore (also ilmenite) with a high-carbon fuel and reductant such as coke, usually with limestone as a flux. Charcoal and anthracite are also used as fuel and reductant.

Pig iron is produced by smelting or iron ore in blast furnaces or by smelting ilmenite in electric furnaces.

Pig iron is supplied in a variety of ingot sizes and weights, ranging from 3 kg up to more than 50 kg.

The vast majority of pig iron is produced and consumed within integrated steel mill complexes. In this context the term "pig iron" is something of a misnomer: within integrated steel mills, blast furnace iron is transferred directly to the steel plant in liquid form, better known as "hot metal" or "blast furnace iron."

The term "pig iron" dates back to the time when hot metal was cast into ingots before being charged to the steel plant. The moulds were laid out in sand beds such that they could be fed from a common runner. The group of moulds resembled a litter of sucking pigs, the ingots being called "pigs" and the runner the "sow."

Merchant Pig Iron

Merchant pig iron is cold pig iron, cast into ingots and sold to third parties as feedstock for the steel and ferrous casting industries.

Merchant pig iron is produced by:

- Dedicated merchant plants - all of whose production is sold to external customers: or.

- Integrated steel mills - with iron that is surplus to their internal requirements and cast into ingots and sold to the merchant market.

Types of Merchant Pig Iron

Merchant pig iron comprises three main types:

- Basic pig iron: used mainly in electric arc steelmaking.

- Foundry pig iron (also known as haematite pig iron): used in mainly in the manufacture of grey iron castings in cupola furnaces.

- High purity pig iron (also known as nodular pig iron): used in the manufacture of ductile (also known as nodular or spheroidal graphite – SG) iron castings.

There are also various sub-types, for example low manganese basic pig iron, semi-nodular pig iron etc.

Composition and Characteristics

Pig iron contains at least 92% Fe and has a very high carbon content, typically 3.5 - 4.5%.

Other constituents are given in the table below:

Typical pig iron characteristics (% by weight)

Pig iron type	C	Si	Mn	S	P
Basic	3.5 - 4.5	≤1.25	≤1.0	≤0.05	0.08-0.15
Foundry	3.5 - 4.1	2.5 - 3.5	0.5 - 1.2	≤0.04	≤0.12
High Purity/Nodular	3.7 - 4.7	0.05 -1.5	≤0.05	≤0.025	≤0.035

Pig iron is supplied in a variety of ingot sizes and weights, ranging from 3 kg up to more than 50 kg.

Benefits in Steelmaking and Ferrous Casting

For further information about pig iron and its advantages in Electric Arc Furnace (EAF) steelmaking and ferrous casting, please see our Fact Sheets:

- Use of Basic Pig Iron in the Electric Arc Furnace (EAF) for steelmaking.

- Use of High Purity Pig Iron for Foundries Producing Ductile Iron Castings.

- Use of Foundry Pig Iron in Grey Iron Castings.

Uses

Traditionally, pig iron was worked into wrought iron in finery forges, later puddling furnaces, and more recently, into steel. In these processes, pig iron is melted and a strong current of air is directed over it while it is stirred or agitated. This causes the dissolved impurities (such as silicon) to be thoroughly oxidized. An intermediate product of puddling is known as *refined pig iron, finers metal*, or *refined iron*.

Pig iron can also be used to produce gray iron. This is achieved by remelting pig iron, often along with substantial quantities of steel and scrap iron, removing undesirable contaminants, adding alloys, and adjusting the carbon content. Some pig iron grades are suitable for producing ductile iron. These are high purity pig irons and depending on the grade of ductile iron being produced these pig irons may be low in the elements silicon, manganese, sulfur and phosphorus. These types of pig iron are used to dilute all the elements (except carbon) in a ductile iron charge which may be harmful to the ductile iron process.

Modern Uses

Until recently, pig iron was typically poured directly out of the bottom of the blast furnace through a trough into a ladle car for transfer to the steel mill in mostly liquid form; in this state, the pig iron was referred to as *hot metal*. The hot metal was then poured into a steelmaking vessel to produce steel, typically an electric arc furnace, induction furnace or basic oxygen furnace, where the excess carbon is burned off and the alloy composition controlled. Earlier processes for this included the finery forge, the puddling furnace, the Bessemer process, and the open hearth furnace.

Modern steel mills and direct-reduction iron plants transfer the molten iron to a ladle for immediate use in the steel making furnaces or cast it into pigs on a pig-casting machine for reuse or resale. Modern pig casting machines produce stick pigs, which break into smaller 4–10 kg piglets at discharge.

Cementite

Cementite consists of iron and carbon compounds combined chemically, having the chemical symbol Fe_3C. It is composed of 93% iron and 7% carbon. This compound is brittle, hard and falls under the ceramic classification.

Cementite plays a vital role in metallurgy. When immersed in a solution of 1%–3% sodium chloride, its corrosion resistance increases significantly. Cementite is also known as iron carbide.

Cementite forms directly through the melt in white cast-iron cases. In the case of carbon steels, it can be formed by either the cooling of austenite or tempering of martensite. As mentioned, cementite consists of 7% carbon; thus in the iron-carbon phase system, the alloy is not cast iron or steel since all its available iron content is in the cementite.

In some cases, cementite combines with ferrite, a byproduct of austenite, in order to build bainite and pearlite or lamellar structures. Since cementite tends to be unstable thermodynamically, it is eventually converted to graphite and ferrite.

Its role in metallurgy is widely appreciated by industries worldwide due to the fact that immersion in sodium chloride solution greatly enhances the corrosion protection properties of this substrate.

Structure of Cementite

Cementite has an orthorhombic unit cell and the common convention is to set the order of the lattice parameters as a=0.50837 nm, b=0.67475 nm and c=0.45165 nm. Note that the order in which the lattice parameters are presented here is consistent with the space group Pnma. There are twelve atoms of iron in the unit cell and four of carbon, as illustrated below. Four of the iron atoms are located on mirror planes whereas the other eight are at general positions (point symmetry 1).

The lattice type is primitive (P). There are n-glide planes normal to the x-axis, at (1/4)x and (3/4)x involving translations of (b/2)+(c/2). There are mirror planes normal to the yy-axis and a-glide planes normal to the z-axis, at heights (1/4)z and (3/4)z with fractional translations of a/2 parallel to the x-axis. The space group symbol is therefore Pnma.

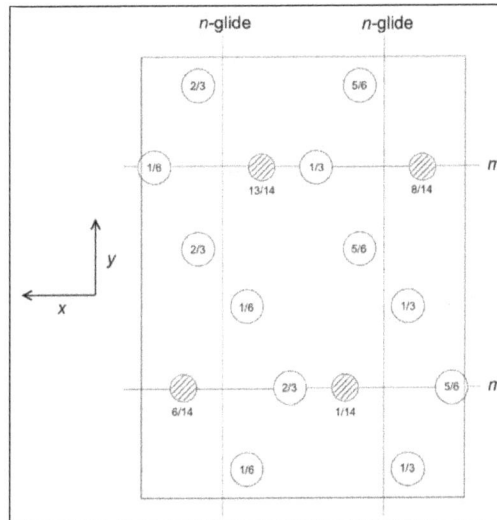

The crystal structure of cementite, consisting of twelve iron atoms (large) and four carbon atoms (small, hatched pattern). The fractional z coordinates of the atoms are marked. Notice that four of the iron atoms are located on mirror planes, whereas the others are at general locations where the only point symmetry is a monad. The pleated layers parallel to (100) are in ABABAB stacking with carbon atoms occupying interstitial positions at the folds within the pleats, with all carbon atoms located on the mirror planes. There are four Fe_3C formula units within a given cell.

Stoichiometry of Cementite

The carbon atoms in cementite are located in interstitial sites; any deficit from the 3:1 Fe:C atom ratio is attributed to vacant interstices that normally are occupied by carbon atoms, as inferred from lattice parameter changes,. The specific volume of cementite that is in equilibrium with ferrite at ambient temperature is found to be greater than that calculated using its measured lattice parameters, indicating those vacant carbon sites, i.e., a deviation from the stoichiometric composition. Similar conclusions have been reached by measuring phase fractions and lattice parameters in rapidly cooled Fe-C alloys containing large carbon concentrations. Indeed, the detailed changes in three lattice parameters of cementite quenched from different temperatures, have been shown to be consistent qualitatively with corresponding parameters calculated using ab initio methods where carbon-specific sites are left unoccupied.

The figure shows the thermodynamically assessed phase boundaries between cementite θ and ferrite α or austenite γ. Cementite has traditionally been depicted as a line compound in phase diagram calculations, but it has been shown that a thermodynamic model that permits its free energy to vary in a manner consistent with experimental data, is able to reproduce the equilibrium γ+θ/θ and α+θ/θ phase boundaries. The fact that ferrite can precipitate from cementite that was equilibrated at elevated temperatures, proves that there is an increase in the amount of carbon within cementite at low temperatures.

Any deviations from stoichiometry must be small, the bond energy between a carbon atom and iron is greater than that between two iron atoms. Therefore, any deficit of carbon would lead to a reduction in cohesion. Any extra carbon beyond the 3:1 Fe:C ratio would need to be accommodated in less-favoured interstices within the cementite lattice.

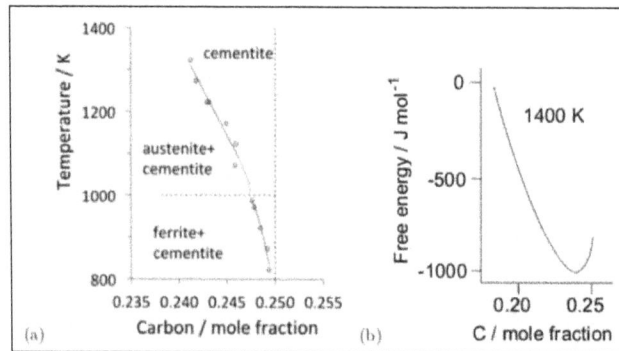

(a) The composition of cementite that is in equilibrium with austenite or with ferrite in an Fe-C alloy. The data are due to Leineweber et al., determined by measuring the lattice parameters of cementite following quenching from the appropriate temperature. (b) Free energy curve of cementite as a function of chemical composition (referred to γ-Fe and graphite). After Gohring et al.

Circumstances can be engineered to make the cementite deviate from the stoichiometric carbon concentration; the decarburisation of pure cementite, which leads to changes in the volume of the unit cell and in the Curie temperature of cementite, is an example. The deviation tends to be small, typically Fe_3C_1-x with $x \cong 0.02$ There are reports that very small particles of cementite in the structure of iron alloys studied by the atom probe technique exhibit deviations from stoichiometry, but these results should be treated with caution because at small size, the surface energy plays a role in determining the composition of the cementite in equilibrium with the surroundings.

The atom probe permits the composition of cementite to be measured directly using time-of-flight mass spectroscopy. There are, nevertheless, difficulties in measuring the carbon concentration of cementite. It has not yet been possible to demonstrate small deviations from stoichiometry using such high-resolution methods. However, using conventional atom probe field ion microscopy, extremely small (4 nm) cementite particles in severely deformed mixtures of ferrite and cementite, have been shown to contain only 16\,at\%\ of carbon, a concentration that recovers to the 25 at% when the mixture is annealed to reduce the defect density and coarsen the cementite. It is argued that the deformation introduces defects such as vacancies into the cementite, leading to the reduction in carbon concentration. However, it is important to note that the particles containing such a large deviation from stoichiometry were not proven to retain the orthorhombic crystal structure.

Thermal Properties

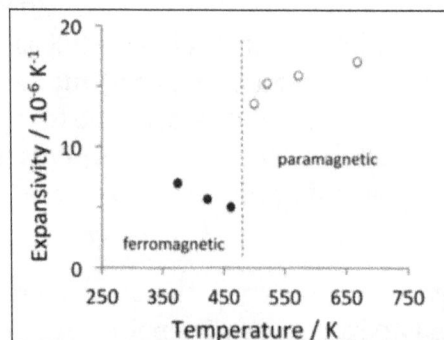

The linear thermal expansion coefficient of polycrystalline cementite as a function of temperature and magnetic state. Adapted using data from Umemoto et al.

The average thermal expansion coefficient of polycrystalline cementite changes from 6.8×10^{-6} K^{-1} to 16.2×10^{-6} K^{-1} as the sample is heated to beyond the Curie temperature.

Shown below are diffraction data for each of the lattice parameters of cementite as a function of temperature. The parameter a is most sensitive to the change from the ferromagnetic to paramagnetic state, with a contraction evident as the temperature is raised within the ferromagnetic range. An increase in the amplitude of thermal vibrations in an anharmonic interatomic potential causes expansion, but the spontaneous magnetisation leads to a contraction, and this latter effect dominates the a parameter below T_C, leading to the observed Invar type effect, although it is known that the analogy with the Invar effect in austenite is tenuous. The orthorhombic structure is preserved through the transition at T_C. It is not clear why the a parameter is particularly affected by the magnetic transition.

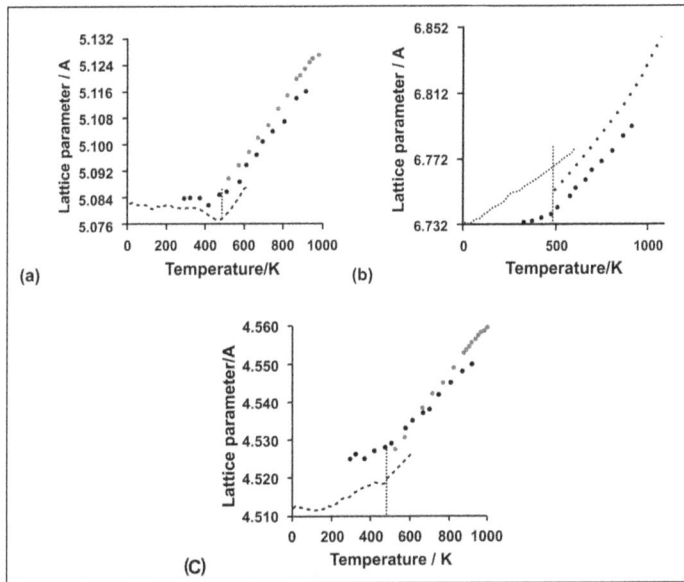

Neutron and X-ray diffraction data on the three lattice parameters a, b and c of cementite as a function of temperature. Data from (small circles with error bars), (filled circles) and (crosses). The dashed line in each case identifies the Curie temperature. The calculated pressure dependencies of the lattice parameters are as follows:

$$\Delta a = 0.0041 \times P, \Delta b = 0.00578 \times P \text{ and } \Delta c = 0.00374 \times P \text{ Å},$$

where the pressure P is in GPa.

Preparation of Cementite

Samples of bulk, pure cementite are difficult to prepare given that cementite in contact with iron is less stable than the corresponding equilibrium between graphite and ferrite. The largest samples have been manufactured by mechanical alloying in experiments. Powders of iron and graphite in the correct stoichiometric ratio, are milled together, resulting in a solid solution, as indicated by very broad ($\cong 15°$) X-ray diffraction peaks in locations typical of body-centred cubic iron. The mechanically alloyed powder was then spark plasma sintered under vacuum at 50 MPa pressure for 300 s at 1173 K, inducing the formation of cementite. The density achieved was 7.5 g cm^{-3}, which is less than the measured value for pure cementite of 7.662 g cm-3, indicating a degree of porosity in the sintered samples.

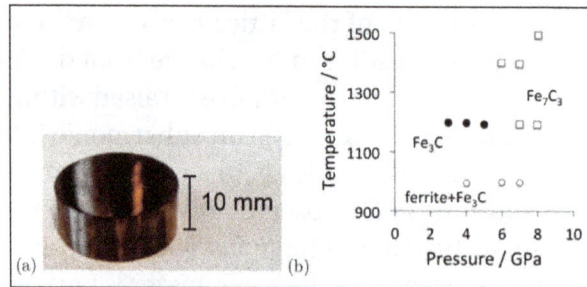

(a) A sample of cementite, courtesy of Professor Minoru Umemoto of Toyohashi University.
(b) Reaction of 80 wt% Fe and 20 wt% graphite for ten minutes at the temperatures and pressures indicated.

The sintering step has been unnecessary in other work where cementite was obtained directly during the milling process. The broadening is caused by strain due primarily to dislocations locked within the powder, indicating a much larger defect density in the samples of the Umemoto study. Carbon prefers to be located at dislocations rather than in cementite; this explains the necessity for the sintering stage in the Umemoto study.

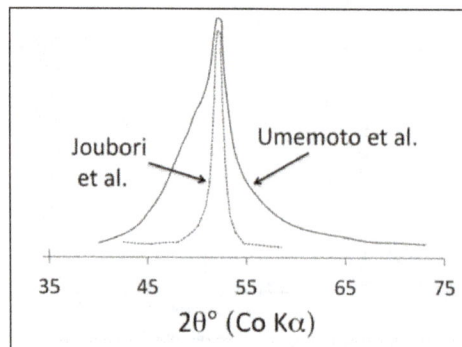

A comparison of the {110}α X-ray peaks from the experiments of Umemoto et al. and Joubouri et al. - the latter has been corrected to the Co Kα wavelength to permit the comparison.

It has been proposed, based on evidence from Mössbauer spectroscopy, that there are intermediate stages between the formation of the solid solution during milling, and that of cementite. The process may first involve transition carbides such as Hägg (Fe_2C) and ε-carbide, followed by cementite. Cementite can be made directly from Hägg carbide through the reaction $Fe + Fe_2C \rightarrow Fe_3C$. Alternatively, powdered cementite can be made by heating Hägg carbide, which is richer in carbon, in a nitrogen stream at 800°C for some 20 min. The resulting sample may contain traces of free iron and amorphous carbon. Cementite also forms when a mixture of iron and graphite heated under a pressure of less than 5 GPa at about 1000°C,. Cementite powders have been made traditionally by electrochemical extraction from steel containing cementite.

A clever method for fabricating a "single crystal" of cementite is to incorporate electrolytically extracted cementite particles into a resin which then is subjected to a 10 Tesla magnetic field for some 24 h with the composite periodically rotated in the field to magnetically align the particles as the resin solidifies. This enabled the magnetocrystalline anisotropy of the cementite to be determined experimentally.

Iron can be converted into cementite by exposing it to a carburising gas mixture, if the activity of carbon relative to graphite is maintained at greater than one. Graphite is deposited preferentially unless the surface of the iron is contaminated with blocking atoms such as sulphur, in which case cementite is precipitated. It has been demonstrated that cementite can be made by carburising magnetite (Fe_3O_4) at 1073 K with carbon monoxide. It is speculated that cementite produced in this manner could be used in an electrical arc furnace to produce iron while at the same time reducing carbon dioxide emissions.

Nanoparticles of cementite can be prepared by the thermal decomposition of $Fe(CO)_5$ (iron pentacarbonyl). These fine particles may be of use in biomedicine for delivery of drugs to specific locations within the body, with the localisation achieved by an external magnetic field. Elemental-iron particles have been proposed for this purpose but they tend to oxidise. Cementite is more corrosion and oxidation resistant. The mechanism of oxidation, i.e. the formation first of Fe_3O_4 followed by Fe_2O_3 remains identical to that of metallic iron. while retaining sufficient ferromagnetism to implement the delivery mechanism. Dispersions of polymer coated cementite nanoparticles have been manufactured by subjecting a gaseous mixture of $C_2H_4/Fe(CO)_5/C_5H_8O_2$ to a continuous wave CO_2 laser pyrolysis.

Cementite powder containing pores about 20 nm in size from an aqueous mixture of iron chloride, colloidal silica and 4,5-dicyanoimidalzole. The dicyanoimidalzole is the source of carbon when the mixture is heated to 700°C to produce the powder of cementite which also contains amorphous silica. The silica is then removed by solution in sodium hydroxide, leaving the porous cementite with a high specific surface area. This cementite was demonstrated to be catalytically active in the decomposition of ammonia into a mixture of hydrogen and nitrogen. Cementite apparently has greater stability under harsh conditions than metallic iron, and is safer with respect to the danger of explosions associated with fine metallic powders. Cementite has in fact been shown to exhibit catalytic activity even in the classical Firscher-Torpsch process for converting gaseous components into hydrocarbon liquids.

Austenite

Austenite is a solid solution of mostly iron and carbon. It has a face-centered cubic crystal structure. Austenite only forms when an iron-based alloy is heated above about 750°C (1382°F) but not above about 1450°C (2642°F). Austenite keeps its form at room temperature when special alloying elements have been added to the iron-based alloy.

Austenite is probably most commonly known for its presence in austenitic stainless steels. Austenite exists in these stainless steels at room temperature because of their high amounts of nickel. Nickel has special properties that promote the formation of austenite in steel and other iron-based alloys. Other elements, such as chromium, deter the formation of austenite and make it more difficult to form in iron-based alloys. Without high amounts of nickel or another austenite-promoting element, an iron-based alloy will typically form pearlite or ferrite when lowered from the temperatures where austenite normally occurs.

Austenitic stainless steels are mostly non-magnetic even though they have high amounts of iron in them. This is because the face-centered cubic arrangement of its atoms is not magnetic, unlike the body-centered cubic structure that ferrite has. The face-centered cubic structure of austenite also allows it to hold higher amounts of carbon than other steel crystal structures.

Austenite was named in honor Sir William Chandler Roberts-Austen, who created the first iron-carbon phase diagram in the 19th century.

Martensite

Martensite is a body-centered tetragonal form of iron in which some carbon is dissolved. Martensite forms during quenching, when the face centered cubic lattice of austenite is distored into the body centered tetragonal structure without the loss of its contained carbon atoms into cementite and ferrite. Instead, the carbon is retained in the iron crystal structure, which is stretched slightly so that it is no longer cubic. Martensite is more or less ferrite supersaturated with carbon. Compare the grain size in the micrograph with tempered martensite.

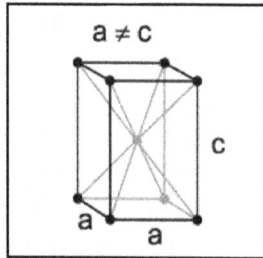

Body Centered Tetragonal Unit Cell   Photomicrograph of Martensite Structure

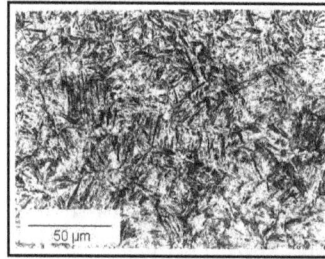

Martensitic Transformation: Mysterious Properties Explained

The difference between austenite and martensite is, in some ways, quite small: while the unit cell of austenite is a perfect cube, in the transformation to martensite this cube is distorted so that it's slightly longer than before in one dimension and shorter in the other two. The mathematical description of the two structures is quite different, for reasons of symmetry, but the chemical bonding remains very similar. Unlike cementite, which has bonding reminiscent of ceramic materials, the hardness of martensite is difficult to explain in chemical terms. The explanation hinges on the crystal's subtle change in dimension, and the speed of the martensitic transformation. Austenite is transformed to martensite on quenching at approximately the speed of sound - too fast for the carbon atoms to come out of solution in the crystal lattice. The resulting distortion of the unit cell results in countless lattice dislocations in each crystal, which consists of millions of unit cells. These dislocations make the crystal structure extremely resistant to shear stress - which means, simply that it can't be easily dented and scratched. Picture the difference between shearing a deck of cards (no dislocations, perfect layers of atoms) and shearing a brick wall (even without the mortar).

Properties

Martensite is formed in carbon steels by the rapid cooling (quenching) of the austenite form of iron at such a high rate that carbon atoms do not have time to diffuse out of the crystal structure in large enough quantities to form cementite (Fe_3C). Austenite is γ-Fe, (gamma-phase iron), a solid solution of iron and alloying elements. As a result of the quenching, the face-centered cubic austenite transforms to a highly strained body-centered tetragonalform called martensite that is supersaturated with carbon. The shear deformations that result produce a large number of dislocations, which is a primary strengthening mechanism of steels. The highest hardness of a pearlitic steel is 400 Brinell whereas martensite can achieve 700 Brinell.

The martensitic reaction begins during cooling when the austenite reaches the martensite start temperature (M_s) and the parent austenite becomes mechanically unstable. As the sample is quenched, an increasingly large percentage of the austenite transforms to martensite until the lower transformation temperature M_f is reached, at which time the transformation is completed.

For a eutectoid steel (0.78% C), between 6 and 10% of austenite, called retained austenite, will remain. The percentage of retained austenite increases from insignificant for less than 0.6% C steel, to 13% retained austenite at 0.95% C and 30–47% retained austenite for a 1.4% carbon steels. A very rapid quench is essential to create martensite.

For a eutectoid carbon steel of thin section, if the quench starting at 750 °C and ending at 450 °C takes place in 0.7 seconds (a rate of 430 °C/s) no pearlite will form and the steel will be martensitic with small amounts of retained austenite.

For steel 0-0.6% carbon the martensite has the appearance of lath, and is called lath martensite. For steel greater than 1% carbon it will form a plate like structure called plate martensite. Between those two percentages, the physical appearance of the grains is a mix of the two. The strength of the martensite is reduced as the amount of retained austenite grows. If the cooling rate is slower than the critical cooling rate, some amount of pearlite will form, starting at the grain boundaries where it will grow into the grains until the M_s temperature is reached then the remaining austenite transforms into martensite at about half the speed of sound in steel.

In certain alloy steels, martensite can also be formed by the working and hence deformation of the steel at temperature, while it is in its austenitic form, by quenching to below M_s and then working by plastic deformations to reductions of cross section area between 20% to 40% of the original. The process produces dislocation densities up to $10^{13}/cm^2$. The great number of dislocations, combined with precipitates that originate and pin the dislocations in place, produces a very hard steel. This property is frequently used in toughened ceramics like yttria-stabilized zirconia and in special steels like TRIP steels. Thus, martensite can be thermally induced or stress induced. One of the differences between the two phases is that martensite has a body-centered tetragonal (BCT) crystal structure, whereas austenite has a face-centered cubic (FCC) structure. The growth of martensite phase requires very little thermal activation energy because the process is a diffusionless transformation, which results in the subtle but rapid rearrangement of atomic positions, and has been known to occur even at cryogenic temperatures. Martensite has a lower density than austenite, so that the martensitic transformation results in a relative change of volume. Of considerably greater importance than the volume change is the shear strain, which has a magnitude of about 0.26 and which determines the shape of the plates of martensite. Martensite is not shown in the equilibrium phase diagram of the iron-carbon system because it is not an equilibrium phase. Equilibrium phases form by slow cooling rates that allow sufficient time for diffusion, whereas martensite is usually formed by very high cooling rates. Since chemical processes (the attainment of equilibrium) accelerate at higher temperature, martensite is easily destroyed by the application of heat. This process is called tempering. In some alloys, the effect is reduced by adding elements such as tungsten that interfere with cementite nucleation, but, more often than not, the nucleation is allowed to proceed to relieve stresses. Since quenching can be difficult to control, many steels are quenched to produce an overabundance of martensite, then tempered to gradually reduce its concentration until the preferred structure for the intended application is achieved. The needle-like microstructure of martensite leads to brittle behavior of the material. Too much martensite leaves steel brittle; too little leaves it soft.

Iron Hydride

Iron hydride (FeHx) is considered as suitable storage for hydrogen in the earth's interior and possibly in the core. Most experimental data on its stability pertain to low pressures (<10 GPa) and temperatures. We studied the reaction of iron with brucite (water) at pressures 75 GPa and temperatures of ~2000 K using the double-side laser-heated diamond-anvil cell. A high-pressure phase of magnetite (Fe_3O_4) (orthorhombic) and iron hydride (double hcp) were found to exist stably under these conditions. The results indicate that at pressures corresponding to the earth's lower mantle, the hydride phase is stable, and that orthorhombic high-magnetite (h-Fe_3O_4) may also be stabilized in lieu of or in addition to magnesiowuestite. The stability of these phases open up the possibility that water (as a component of a fluid phase or hydrous solids) may be present not only in the mantle but also in the core (as dissolved hydride and oxide), which helps melting and dynamic movements. The core may have been the reservoir of oceans of fluid. A percent of water (by weight) in the core is equivalent to about 10 times the water in all the oceans. The dissolved water components in the core would depress the melt-ing temperature of iron (or iron–nickel alloy) significantly, reduce the density and effectively promote convection.

The Fe–H System

Iron belongs to the group VI–VIII transition metals, of which only Pd forms hydrides at low hydrogen pressures. The hydrides formed at high pressures in the systems other than Fe–H were shown to have close-packed metal sublattices with an fcc (γ) or hcp (ε) structure, in which hydrogen occupies octahedral interstitial positions. Most of these hydrides exist in wide composition ranges and can be considered as solid solutions of hydrogen distributed over the interstices either randomly or with a superstructure order. Hydrogen induces a rather large volume expansion $\Delta V_a(x)$ of the host metal that reaches 2–3 Å³ per metal atom at x = 1 and can serve as an indicator of hydride formation.

Iron forms an ε' hydride with a double hcp metal lattice and $x \approx 1$, an fcc γ hydride of unknown composition and also a metastable intermediate hcp ε phase with $x \approx 0.4$. The ε'-FeH hydride is a ferromagnet with a spontaneous magnetization of $\sigma_0 \approx 2.2\ \mu_B$ per iron atom at T = 0 K and the Curie temperature much exceeding 300 K. The ε-FeH$_{0.42}$ hydride is paramagnetic down to 4.2 K. The magnetic properties of the γ hydride are not known yet. The hydrides rapidly decompose at ambient pressure on heating above 150 K.

The T −P diagram of phase transformations of iron placed in an atmosphere of gaseous hydrogen is shown in figure. It was constructed from the positions of anomalies of the pressure and temperature dependences of the electrical resistance and magnetic permeability determined with an accuracy of ± 0.3 GPa and ± 15 °C. The lines of phase transformations divide the T − P plane of the diagram into three regions: α, ε and γ. The co-ordinates of the triple point are T ≈ 280 °C and P ≈ 5 GPa.

According to the phase rule, only single-phase fields are possible in the equilibrium T −P diagrams of binary Me−H systems, but the composition of every phase can vary continuously with T and P within the corresponding field. At temperatures up to 350 °C, the composition of the ε' -hydride is close to FeH independent of the pressure, while at higher temperatures a decrease in V_a indicates a gradual decrease of the hydrogen content. The $\alpha \rightarrow \varepsilon'$ transition in the Fe−H system is very sluggish, and a noticeable amount of non-reacted α-Fe was observed even in Fe−H samples loaded with hydrogen at a pressure as high as 9 GPa at room temperature and at 350 °C. The considerable hysteresis of the $\alpha \leftrightarrow \varepsilon'$ transformation is caused by elastic stresses, which are close to the yield stress and appear to be due to the sudden increase in sample volume by about 16% upon hydride formation. The curve of thermodynamic equilibrium between the α- and ε' -phase therefore should be close to the $\varepsilon' \rightarrow \alpha$ curve of hydride decomposition, since removal of hydrogen leads to the relaxation of accumulated elastic stresses. This is characteristic of phase transformations in all metal− hydrogen systems at high pressures, and the ε → α curve in figure is likely to well represent the phase equilibrium.

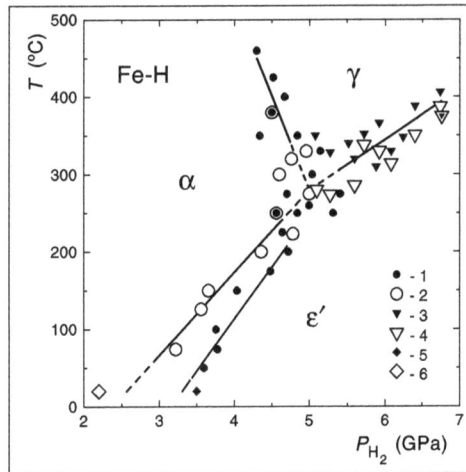

In figure, T −P phase diagram of the Fe−H system. 1, 2 are the midpoints of anomalies of the electrical resistance isotherms measured at increasing and decreasing pressure, respectively; 3, 4 are the midpoints of anomalies of the isobars of electrical resistance and magnetic permeability measured at increasing and decreasing temperature, respectively; 5, 6 are the formation and decomposition pressure of ε'-FeH at room temperature from measurements in diamond anvil cells. α is the dilute hydrogen solid solution in bcc α -Fe with x < 0.01; ε is the approximately stoichiometric ε -FeH hydride with a double hcp metal lattice; γ is the hydride with an fcc metal lattice and as yet unknown composition.

In situ x-ray studies at a hydrogen pressure of 6 GPa and elevated temperatures showed that the γ -phase has an fcc metal lattice and nearly the same volume V_a per metal atom as the ε' hydride under the same conditions, so the γ-phase was assumed to be an iron hydride as well. Further x-ray investigations allowed the Japanese group to outline the T − P diagram of the Fe−H system over a wide range of temperatures and pressures . Its comparison with the T −P diagram of Fe in inert media figure shows that the γ-phase can be considered as a solid solution of hydrogen in the allotropic γ modification of Fe. However, all efforts to quench the γ-phase under pressure for further studies at ambient pressure were unsuccessful, and its composition and physical properties still remain unknown.

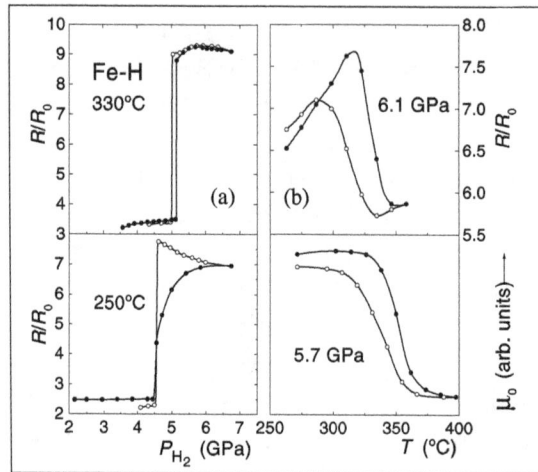

Representative isotherms of the electrical resistance (a) and isobars of the electrical resistance and magnetic permeability (b) of iron measured at increasing (solid circles) and decreasing (open circles) pressure or temperature in a hydrogen atmosphere. Ro is the initial resistance of the sample at ambient temperature and pressure.

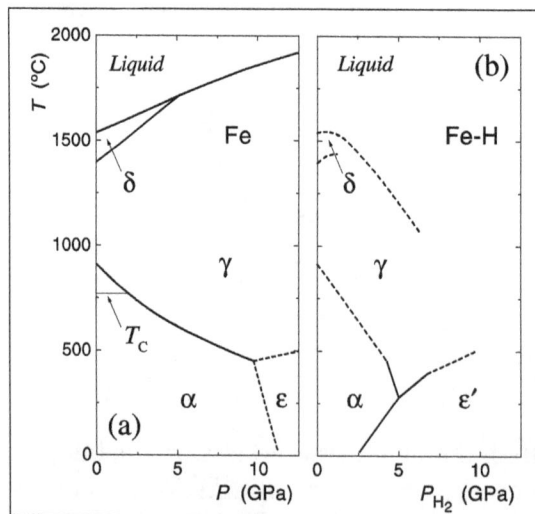

T−P phase diagrams of Fe in inert media (a) and in a hydrogen atmosphere (b). The solid lines in (b) represent the phase boundaries from shown in figure; the dashed lines show tentative boundaries in the extended T −P region according to the in situ x-ray diffraction studies of . The letters in (a) and (b) mark the fields of phases with different metal lattices: α, δ = bcc, γ = fcc, ε = hcp, ε' = dhcp. T_c is the Curie temperature of ferromagnetic α-Fe.

X-ray diffraction also showed that the $\alpha \rightarrow \varepsilon'$ transition at elevated temperatures is a complex process involving formation of an intermediate ε-phase with an hcp metal lattice and a volume per metal atom, V_a, lying somewhere in between those for pure hcp ε-Fe and ε'-FeH. Similar to ε'-FeH, the ε-phase can be quenched at high pressure and studied at ambient pressure.

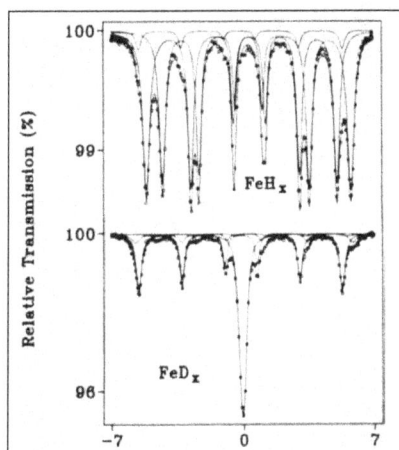

In figure above, Representative [57]Fe Mossbauer spectra of Fe–H and Fe–D samples measured at 4.2 K¨ with a [57]Co:Rh source also at 4.2 K. The spectrum of the Fe–H sample consists of two magnetic patterns of ε'-FeH with nearly equal intensities (53 and 43%) and a weak (4%) sextet of α-Fe. The main components of the Fe–D spectrum are the non-magnetic single-line pattern of the ε-phase (41%) and the magnetic sextet of α-Fe (44%).

Table: The main components of the [57]Fe Mossbauer spectra shown in figure. S is the isomer shift with respect to the source of [57]Co in rhodium at 4.2 K; E_Q the effective electric quadrupole interaction; B_{hf} is the magnetic hyperfine field; RI is the relative intensity of the individual components.

Sample	S (mm s⁻¹)	E_Q (mm s⁻¹)	B_{hf} (T)	RI (%)	Assignment
Fe–H	+0.364(5)	−0.053(5)	33.8(1)	53(1)	ε'-FeH, site 'h
	+0.370(5)	+0.030(5)	28.8(1)	43(1)	ε'-FeH, site 'c'
	−0.122(5)	0.0	33.8(1)	4(1)	α-Fe
Fe–D	−0.037(5)	----	----	41(2)	ε-FeHx , x ~ 0.3
	−0.128(5)	0.0	33.7(1)	44(1)	α-Fe

Figure and table present the results of a Mossbauer investigation of the Fe–H and Fe–D samples synthesized at 350 °C and H_2 or D_2 pressures of about 9 GPa. The Fe–H sample was nearly pure ε'-hydride, while the Fe–D sample contained the maximum amount of the ε-phase obtained in that work. (No evident relationship has been found so far between the amount of the ε-phase in the sample and the conditions of synthesis within the ε' field of the T –P diagram; the content of the ε-phase was, however, usually larger in samples loaded with deuterium than in samples loaded with hydrogen.)

In the dhcp iron lattice of ε'-FeH, there are two crystallographically inequivalent iron sites, i.e. 'hexagonal' and 'cubic' ones, arranged in close-packed layers that lie, respectively, between adjacent layers stacked in the same way or stacked differently. Accordingly, as was first observed in, the [57]Fe Mossbauer spectrum of ferromagnetic ε'-FeH consists of two magnetic six-line patterns, which have practically the same isomer shift. The pattern with the larger hyperfine field B^{hf} has a negative effective electric quadrupole interaction ΔEQ, while the pattern with the smaller B_{hf} has a positive ΔE_Q. The pattern with the negative ΔE_Q was ascribed to Fe atoms on the 'hexagonal'sites, because the electric quadrupole interaction is negative for Fe substituting Co in ferromagnetic cobalt hydrides with an hcp metal lattice composed of equally stacked 'hexagonal' close-packed layers (the Co hydrides were studied in the same work). The Mossbauer spectra of hydrides of Fe–Cr alloys discussed later in the present work lend further credence to this attribution. The ideal dhcp stacking sequence is. hchchc with equal numbers of 'h' and 'c' layers. Interestingly, the area under the 'h' pattern is about 20% larger than that under the 'c' pattern. The difference was attributed to the presence of stacking faults in the dhcp iron lattice of ε' -FeH leading to the increase in the number of 'h' layers.

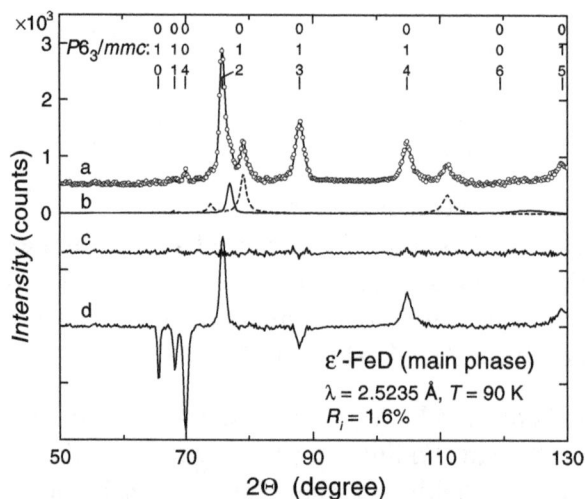

In figure above, Neutron diffraction pattern of an Fe–D sample (circles) and results of its Rietveld analysis (lines) involving ε' -FeD (the main phase) and ε-FeD$_{0.42}$ and α-Fe. The final values of the fit parameters are listed in table (a) Calculated profile shown as a solid curve; (b) calculated contributions from ε-FeD$_{0.42}$ (dashed curve) and from α-Fe (solid curve); (c) difference between the experimental (circles) and calculated (curve (a)) spectra; (d) difference between the experimental spectrum and that calculated for deuterium randomly occupying half of the tetrahedral interstitial sites in the dhcp lattice of ε' -FeD, the other fit parameters being the same as in table. The indices in the upper part of the figure refer to ε' -FeD.

The single-line pattern from the ε-phase in the Fe–D sample shown in figure suggests that this phase is not magnetically ordered at 4.2 K. The isomer shift of the pattern corresponds to the hydrogen content of the ε-phase of x $\approx$ 0.3.

One Fe–H sample consisting nearly entirely of ε' hydride and two Fe–D samples containing the minimum and the maximum amounts of the ε-phase were studied by neutron diffraction. The samples contained a few per cent of non-reacted α-Fe. The structures of all phases were refined using Rietveld profile analysis. The diffraction pattern of one sample and the results of its profile fit are illustrated by figure. The proposed crystal structures of the ε ε' - and ε-phases are schematically shown in figure. The refined parameters of these phases are listed in table.

The profile fit showed that the composition of the ε-phase is ε-FeD$_{0.42(4)}$ and that deuterium randomly occupies octahedral interstices in its hcp metal lattice (open circles in figure). The octahedral hydrogen coordination is characteristic of all hcp and fcc hydrides of the group VI–VIII transition metals studied so far. Unlike ε-FeD$_{0.42}$(4), however, the two other known ε hydrides with x $\approx$ 1/2, CoD$_{0.5}$ and TcH$_{0.45}$, form layered superstructures of the anti-CdI$_2$ type with D or H atoms preferentially occupying every second basal layer of octahedrons.

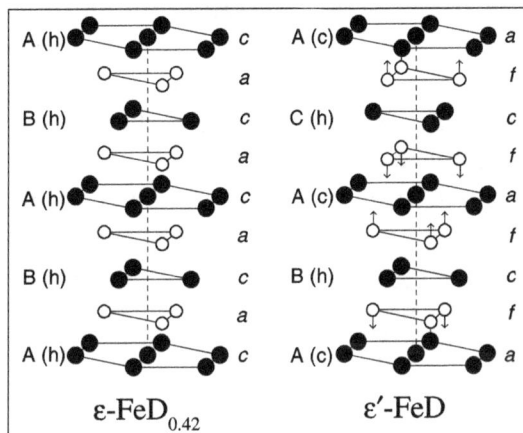

ε-FeD$_{0.42}$ ε'-FeD

In figure above, Crystal structures of ε' -FeD and of ε-FeD$_{0.42}$. The letters a, c and f mark layers of equivalent positions in the P6$_3$/mmc space group originating, respectively, from positions 2a, 2c and 4f with Z = 0.882 $\approx$ 7/8. The solid circles show regular positions of Fe atoms, open circles of D atoms. The arrows indicate the directions of the displacements of D atoms from the centres of the octahedral interstices in ε' -FeD. The letters A, B and C represent the standard notation for close-packed layers in the hcp and dhcp structures; (h) and (c) indicate layers of 'hexagonal' and 'cubic' metal sites.

Table: Positional parameters (X, Y, Z) and site occupancies for iron hydride and deuterides according to the Rietveld profile analysis of the neutron diffraction data collected at 90 K using the D1B diffractometer at ILL, Grenoble. M is the number of formula units per unit cell; N is the number of atoms per unit cell.

Phase	Atom	Site	X	Y	Z	Occupancy	N
ε'-FeH, P6$_3$/mmc, M = 4	Fe	2a	0	0	0	1.000	2.00
a = 2.679 Å, c = 8.77 Å,	Fe	2c	1/3	2/3	1/4	0.935	1.87
c/a = 2 × 1.637	H	4f	1/3	2/3	7/8	0.935	3.74
	Fe	2d	1/3	2/3	3/4	0.065	0.13
	H	4f	1/3	2/3	1/8	0.065	0.26
ε'-FeD, P6$_3$/mmc, M = 4	Fe	2a	0	0	0	1.000	2.00
a = 2.668 Å, c = 8.75 Å,	Fe	2c	1/3	2/3	1/4	0.845	1.69
c/a = 2 × 1.640	D	4f	1/3	2/3	0.882	0.845	3.38
	Fe	2d	1/3	2/3	3/4	0.155	0.31
	D	4f	1/3	2/3	0.118	0.155	0.62
ε-FeD$_{0.42(4)}$; P6$_3$/mmc, M = 2	Fe	2c	1/3	2/3	1/4	1.00	2.00
a = 2.583 Å, c = 4.176 Å	D	2a	2a	0	0	0.42	0.84
c/a = 1.617							

In the ε'-phase, hydrogen or deuterium also fill octahedral interstitial positions, the alternative tetrahedral model being completely inadequate, as is shown by the difference curve (d) in figure. The excessive number of iron 'h' layers due to stacking faults resulted in a significant decrease of the intensities of those diffraction lines of the dhcp structure that are 'superstructural' for the corresponding hcp lattice. To incorporate stacking faults into the calculation scheme for the space group $P6_3/mmc$, partial occupation of 'defect' d sites by Fe atoms and f sites with $Z \approx 1/8$ by H or D atoms was allowed along with occupation of 'regular' sites. As seen from table, the obtained number N of H or D atoms on 'defect' sites is twice as large as that of Fe for both ε'-FeH and ε'-FeD. This can be regarded as a successful self-test of the model, since a replacement of a c layer of iron atoms by a d layer, which is just a rotation of the layer through an angle of 60°, will also rotate two adjacent hydrogen (deuterium) layers from regular to defect f positions.

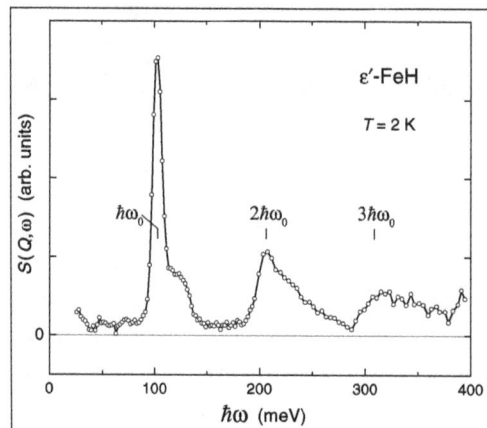

The dynamical structure factor S(Q, ω) of ε'-FeH as a function of the energy loss $\hbar\omega$ of the inelastically scattered neutrons measured at 2 K. The main optical hydrogen peak is at $\hbar_{\omega 0}$ = 103 meV.

The neutron scattering amplitudes of D and H atoms are significantly different, and, in contrast to the neutron diffraction pattern of ε'-FeH, the pattern of ε'-FeD exhibits intensive peaks at large angles. This allowed a more accurate analysis of possible displacements of D atoms from the centres of octahedral interstices along the c-axis. The optimization procedure yielded $\delta Z = 0.007 \times c \approx 0.06$ Å. Though the effect was largely within the error limits, such displacements are very likely on physical grounds, as they increase the distance between deuterium atoms in equally stacked octahedral layers, which is the shortest distance between any deuterium atoms in the ε'-FeD structure. A tendency to increase the D–D distance is expected in view of the strong long-range repulsive interaction between hydrogen atoms, which is one of the main factors governing the formation of transition-metal hydrides.

The spectrum of optical hydrogen vibrations in ε'-FeH has been studied by inelastic neutron scattering (INS) using the IN1 BeF spectrometer at ILL, Grenoble. The spectrum is shown in figure and looks similar to the INS spectra of hydrides of all other 3d and 4d metals of groups VI–VIII studied so far. The first, fundamental band of optical H vibrations consists of a strong peak centred at $\hbar\omega_0$ = 103 meV with a broad shoulder towards higher energies. Based on results for palladium hydride, the main peak is usually ascribed to nearly nondispersive transverse optical modes, while the shoulder is assumed to arise from longitudinal optical modes, which show significant dispersion due to long-range repulsive H–H interactions. The second and the third optical H band have a smoother intensity distribution and appear at energies approximately two and three times the energy of the fundamental band, respectively. As a function of the hydrogen–metal distance R, the value of $\hbar\omega_0$ for ε'-FeH agrees well with the approximately linear dependence $\hbar\omega_0$ (R) for ε and γ hydrides of 3d metals.

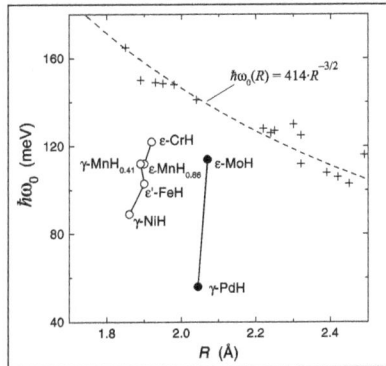

Energy of the main optical hydrogen peak, $\hbar\omega_0$, versus the shortest hydrogen–metal distance R for various dihydrides with a fluorite-type structure (crosses) and for monohydrides of 3d metals (open circles) and 4d metals (solid circles) for references). The dashed curve is a least-squares fit to the data for the dihydrides.

Hydrides of Iron Alloys

The Rigid-d-band Model

Concentration dependences of the spontaneous magnetization σ_0 at T = 0 K (for ferromagnets) and of the Neel temperature T_N (for antiferromagnets) of fcc (γ) alloys of 3d metals which are close neighbours in the periodic table are well described by the rigid-band model and can be represented as unique functions of the average number N^e of external (3d + 4s) electrons per atom of the alloy (the so-called Slater–Pauling curves). These dependences are plotted by thin solid curves in figure. Our studies of γ hydrides of fcc Ni-based and Fe-based alloys showed that the magnetic properties changed with increasing hydrogen content as if the hydrogen were merely a donor of a fractional number of $\eta \approx 0.5$ electrons per H atom in the otherwise unchanged metal d band. This approximation, which we call the rigid-dband model for brevity, is a rather straightforward consequence of the available ab initio band structure calculations, and its applicability to alloys with different types of band structure has been thoroughly discussed in. In particular, if the properties of the alloy obey the rigidband model, the properties of its γ hydrides considered as functions of the effective electron concentration N^e (x) = N^e (0) + ηx are described by the thin curves of figure.

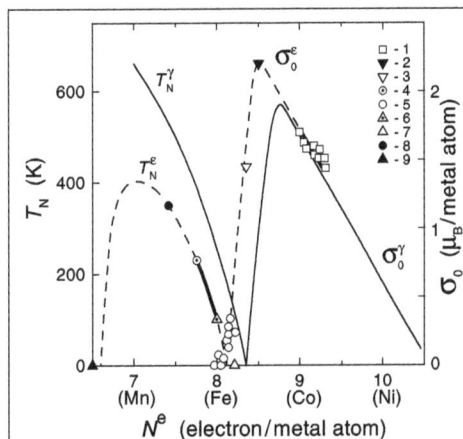

In figure, Slater–Pauling curves for fcc (γ) alloys (experiment–thin solid curves) and hcp (ε) alloys (experiment–two sections of thick solid curves; an estimation–dashed curves) of 3d metals that are nearest neighbours in the periodic table. The symbols show experimental data for hydrides presented as a function of the effective electron concentration, Ne (x) = Ne (o) + ηx, with η = 0.5 electrons per H atom: 1– σ_0 of ε-CoH$_x$ solid solutions; 2–σo of   -FeH; 3–σ_0 of Fe$_{0.947}$Cr$_{0.053}$H$_{0.92}$ with a 9R-type metal lattice; 4– T$_N$ of ε-Fe; 5– σ_0 of ε-FeH$_{0.42}$; 6–T$_N$ of ε-Fe$_{0.776}$Mn$_{0.224}$; 7–σ_0 of ε-Fe$_{0.776}$Mn$_{0.224}$H$_x$ solid solutions; 8– TN of ε-Mn$_{0.83}$; 9–T$_N$ of ε-CrH.

Due to the relatively narrow intervals of mutual solubility of 3d metals in hexagonal phases, the concentration dependences of their magnetic properties are known to a much lesser extent than those of the fcc alloys (two segments of thick solid line in figure). However, if the available experimental points for hexagonal hydrides are added to the graph as a function of N^e (x) with η = 0.5 electrons per H atom, the properties of both hexagonal metals and hydrides can be described by the same curves. These curves are similar to those for the γ alloys and are likely to represent the Slater–Pauling plot for the (hypothetical) hexagonal alloys.

Hydrides of Fe–Cr Alloys

Three bcc (α) Fe–Cr alloys containing 50, 25 and 5.3 at.% Cr were loaded with hydrogen at 325 °C and various H2 pressures up to 7 GPa, and the resulting hydrides were studied in a metastable state at ambient pressure by x-ray diffraction and magnetization measurements. The hydrides of alloys containing 25 and 50 at.% Cr were found to have an hcp metal lattice like CrH. At a Cr content of 5.3 at.%, the hexagonal close packed structure has been found to contain stacking faults ordering at least partially into a 9R structure, which may be considered as a transitional state between the hcp metal lattice of the Cr-rich alloys and the dhcp structure of the hydride of pure iron. The hydrides of Fe–Cr alloys with hydrogen-to-metal ratios close to unity have been found to be ferromagnetic or at least contain ferromagnetic fractions up to Cr contents of 25 at.%.

In the present work, [57]Fe Mossbauer experiments were performed at 4.2 K on hydrides of the same three Fe–Cr alloys, which were loaded with hydrogen at 325 °C and pressures of 2.8 and 7 GPa. The results of the Mossbauer experiments are largely in agreement with the magnetic properties of the hydrides of Fe–Cr alloys, but they yield additional information on the individual lattice sites and phases.

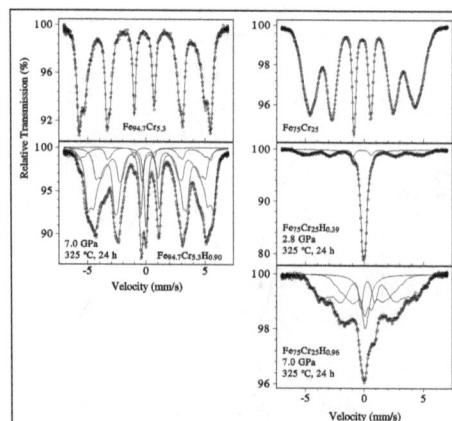

Mossbauer spectra of Fe–Cr alloys and t heir hydrides measured at 4.2 K with a [57]Co:Rh source also at 4.2 K.

The Mossbauer pattern of α-$Fe_{0.947}Cr_{0.053}$ resembles those reported by Dubiel and Inden for Fe–Cr alloys of similar composition and was fitted by the model suggested by these authors, which assumes that m nearest and n next-nearest Cr neighbours reduce the hyperfine field at the Fe by $m\,\Delta\,H_1$ and $n\,\Delta\,H_2$, respectively, and that the distribution of Cr atoms around the iron is binomial. The spectrum of $Fet_{0.947}Cr_{0.053}H_{0.90}$ can largely be described as a superposition of two such patterns, one for the 'h' sites and one for the 'c' sites in the, at least approximate, 9R structure of the hydride. With 33.3 and 29.1 T, the fields for iron in $Fe_{0.947}Cr_{0.053}H_{0.90}$ with m = n = 0 are nearly the same as those for the two sites in the dhcp iron hydride. In $Fe_{0.947}Cr_{0.053}H_{0.90}$, however, the component with the bigger hyperfine field is about twice as strong as that with the smaller field.

In the 9R structure, there are twice as many 'h' sites as 'c' sites. The Mossbauer pattern ¨ with the higher field must thus be attributed to the 'h' sites, and that with the lower field to the 'c' sites. This confirms the previous attribution of the Mossbauer component with the higher ¨ field in dhcp iron hydride to the 'h' sites.

The isomer shift of both Mossbauer components in $Fe_{0.947}Cr_{0.053}H_{0.90}$ with respect to α-$Fe_{0.947}Cr_{0.053}$ is 0.49 mm s^{-1}, which is practically the same as the shift between ε'-FeH and α-Fe.

The spectrum of $Fe_{0.947}Cr_{0.053}H_{0.90}$ also contains a minor magnetically split contribution of α-$Fe_{0.947}Cr_{0.053}$ and a weak single line with a width of 0.48 mm s^{-1} and a shift of 0.14 mm s^{-1} with respect to α-$Fe_{0.947}Cr_{0.053}$. The presence of the non-magnetic component in the spectrum of $Fe_{0.947}Cr_{0.053}H_{0.90}$ isreminiscent of the single line observed in some hydrides and deuterides of pure iron, which has a similar isomer shift with respect to α-Fe. This isomer shift indicates that the hydrogen content of the non-magnetic phase is smaller than that of the magnetic hydride. The absence of ferromagnetism at low hydrogen contents is in agreement with the rigid-d-band model, according to which hexagonal or cubic close-packed Fe (N^e < 8 electrons per atom) and Fe–Cr alloys (Ne < 8 electrons per atom) must contain a certain amount of hydrogen to become ferromagnetic.

$Fe_{0.75}Cr_{0.25}H_{0.39}$ yields mainly a single line with an isomer shift of 0.08 mm s^{-1} with respect to hydrogen-free α-$Fe_{0.947}Cr_{0.053}$. The weak magnetically split fraction in this spectrum represents a minor residue of the hydrogen-free $Fe_{0.75}Cr_{0.25}$ alloy according to its isomer shift and hyperfine field. Assuming that this fraction of the sample contains no hydrogen at all, one concludes that the hydride giving rise to the unsplit Mossbauer line has a hydrogen-to-metal ratio near x = 0.6, for which the hydrides are, indeed, expected to be nonmagnetic according to the rigid-d-band model, because this composition corresponds to only Ne (0.6) ≈ 7.8 electrons per metal atom. According to the rigid-d-band model, the hydrides should, however, become ferromagnetic at higher x. This is borne out by $Fe_{0.75}Cr_{0.25}H_{0.96}$, which gives rise to a broad, unresolved magnetic Mossbauer pattern, with a weak (12%) single-line contribution. This non-magnetic phase may be due to inhomogeneities in the hydrogen distribution.

$Fe_{0.50}Cr_{0.50}H_x$ hydrides with x = 0.81 and 0.99 are non-magnetic and give rise to somewhat broadened single Mossbauer lines with isomer shifts of 0.23 and 0.35 mm s^{-1} with respect to hydrogen-free α-$Fe_{0.947}Cr_{0.053}$. It is again in agreement with the rigid-d-band model that at a Cr content of 50% the hcp hydrides should no longer become ferromagnetic even at x = 1.

Solid Solutions of Hydrogen and Deuterium in the fcc Ni$_{0.8}$Fe$_{0.2}$ Alloy

Nickel hydride is formed via an isomorphous $\gamma_1 \rightarrow \gamma_2$ transition that terminates in a critical point of the liquid–vapour type at T$_{cr} \approx$ 390 ∘C and a hydrogen pressure of P$_{cr} \approx$ 1.75 GPa. Iron and nickel form an fcc disordered substitutional solid solution within a wide concentration range. Alloying of fcc Ni with Fe decreases $_{Tcr}$, and the solubility of hydrogen in the Ni–Fe alloys containing more than about 15 at.% Fe is a continuous function of pressure at T > 20 ˚C.

Hydrogen is known to form continuous γ -solutions at room temperature with many Nibased and Pd-based alloys. X-ray diffraction studies showed that a characteristic feature of most such solutions, including Ni–Fe–H, is a significant decrease in the slope of the $\Delta V_a(x)$ dependence at H-to-metal atomic ratios x $\geq$ 0.7–0.8. The effect has already been debated for some decades and various explanations have been proposed, such as filling of the d band, formation of vacancies in the metal sublattice and even metallization of the hydrogen sublattice. However, none of the explanations has had adequate experimental grounds so far.

To learn more about the origin of the non-linear $\Delta V_a(x)$ dependence and about the hydrogen distribution over interstitial sites in the metal lattice, we carried out an x-ray, neutron diffraction and ^{57}Fe Mossbauer investigation of solid solutions of hydrogen and deuterium in the fcc Ni–Fe alloy containing 20 at.% Fe.

X-ray and Neutron Diffraction Studies

Foils of a disordered substitutional Ni$_{0.8}$Fe$_{0.2}$ alloy were loaded with hydrogen or deuterium by a 24 h exposure to a H$_2$ or D$_2$ atmosphere at a given pressure and at temperatures of 325 or 250 ˚C, followed by cooling to 150 K under pressure in order to avoid H or D losses from the samples when the pressure was released. The synthesis pressure was measured with an accuracy of ±0.3 GPa, the temperature with an accuracy of ±7 ˚C. The mean hydrogen or deuterium content of the obtained samples was determined with an accuracy of 2% by hot extraction at temperatures up to 500˚C. The resulting pressure–composition isotherms are shown in figure.

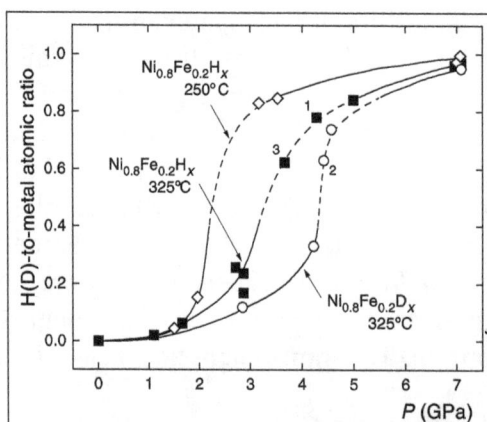

Isothermal absorption relationships between H$_2$ or D$_2$ gas pressure and H or D content x, respectively, of a Ni$_{0.8}$Fe$_{0.2}$ alloy. Numbers 1, 2 and 3 mark the points for the same samples in figures. Sample 3 was prepared differently from the others.

The x-ray diffraction studies showed that all the Ni$_{0.8}$Fe$_{0.2}$Hx and Ni$_{0.8}$Fe$_{0.2}$Dx foils were single phase and had an fcc metal lattice. The obtained values of $\Delta V_a(x)$ are shown by open symbols in figure. Similar to analogous dependences found in the literature, the $\Delta V_a(x)$ dependence thus obtained can be approximated with two straight lines, one below and one above x = 0.7. These lines are shown dashed in figure and their slopes, $\beta = (\partial/\partial_x)\,\Delta V_a(x) = 2.67$ and 0.93 Å3 per metal atom per H(D) atom, are close to those for other γ -solutions of hydrogen in nickel and palladium alloys.

Some of the $Ni_{0.8}Fe_{0.2}Hx$ and $Ni_{0.8}Fe_{0.2}Dx$ samples were also studied by neutron diffraction at 124 K with the D20 diffractometer at ILL, Grenoble. To avoid texture effects, the samples had been previously powdered in an agate mortar under liquid nitrogen. Typical results are shown in figure. The $Ni_{0.8}Fe_{0.2}Hx$ solid solutions are ferromagnets and their spontaneous magnetization at 124 K decreases approximately linearly from 1.1 μ_B per metal atom at x = 0 to $0.6\mu_B$ per metal atom at x = 1 . To reduce the number of fitting parameters, the magnetic contribution to the scattering intensity was calculated using the values of spontaneous magnetization given by this experimental linear dependence. A model assuming H or D atoms randomly distributed over the centres of octahedral interstitial positions in the fcc metal lattice gave a good profile fit of each diffraction pattern, while an occupation of tetrahedral interstices by hydrogen was found to be inconsistent with the experiments.

The values of $\Delta V_a(x)$ with both ΔV_a and x determined from the profile fit of the neutron diffraction data are shown in figure by solid symbols. Two samples, with x ≈ 0.8 and 0.63 (the corresponding points in figure are labelled 1 and 2), showed a significant decrease in the lattice parameters after powdering. X-ray studies of the powdered samples gave the same new values of the lattice parameters as neutron diffraction.

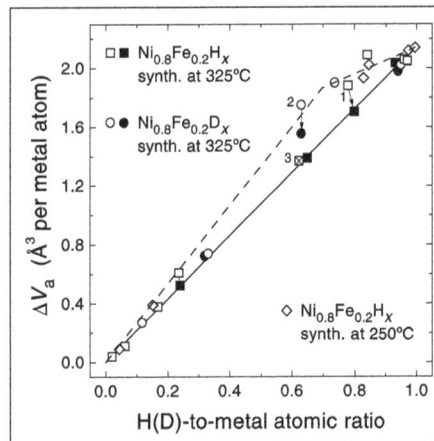

In figure above, The dependence of $\Delta V_a = [a^3(x) - a^3(0)]/4$ on x for fcc $Ni_{0.8}Fe_{0.2}Hx$ and $Ni_{0.8}Fe_{0.2}Dx$ solid solutions. The open symbols show the results of x-ray diffraction studies on foils at 100 K; the hydrogen or deuterium contents were obtained by hot extraction. The solid symbols refer to neutron diffraction measurements on powdered samples at 124 K, the H or D contents resulting from the structure refinement. The arrows connect the points for the same samples before and after the powdering. a(0) = 3.539 Å.

As no hydrogen losses should occur in the course of the powdering under liquid nitrogen, the observed decrease in the lattice parameters suggests that the samples with x ≈ 0.8 and 0.63 were inhomogeneous. In fact, with Fe Kα radiation used, the x-ray diffraction patterns from the 250 μm thick foils were formed only by surface layers with a thickness of a few micrometres. Therefore, about 98% of the sample volume remained unexplored. In contrast, the neutron measurements gave diffraction patterns averaged over the entire sample volume. The x-ray diffraction on the powdered samples would also give effectively volume-averaged patterns. The results of the diffraction studies of the samples with x ≈ 0.8 and 0.63 before and after powdering can be understood if the surface layers of the foils of these samples are assumed to have had a higher hydrogen concentration and a larger lattice parameter than the inner parts.

The neutron diffraction pattern of the $Ni_{0.8}Fe_{0.2}H_{0.8}$ sample showed no traces of the hydrogen-rich surface phase. The lines of the main phase were narrow and the profile fit converged to a good ratio $R_p/R_{ex} = 1.7/1.9$ of the obtained and expected profile factors, indicating a good homogeneity of the specimen. Since a 5% admixture of a hydrogen-rich phase would be clearly seen in the diffraction pattern, we can roughly estimate that such a phase could form a layer no thicker than 15 μm on the surface of the 250 μm thick disc, while the remaining >95% disc volume consisted of a homogeneous phase with a lower hydrogen concentration. Thus, the hydrogenation of the sample with x ≈ 0.8 must have been controlled by processes in a thin surface layer, while the diffusion rate in the bulk was high enough to provide a uniform hydrogen distribution.

The $Ni_{0.8}Fe_{0.2}D_{0.63}$ sample appeared inhomogeneous throughout. The irregular profile of its broadened neutron diffraction lines could not be fitted accurately at high diffraction angles, which resulted in $R_p/R_{ex} = 9.3/4.5$ and a large uncertainty in the x value. The observed inhomogeneity suggests a low rate of hydrogen diffusion in the bulk of the sample with x ≈ 0.63.

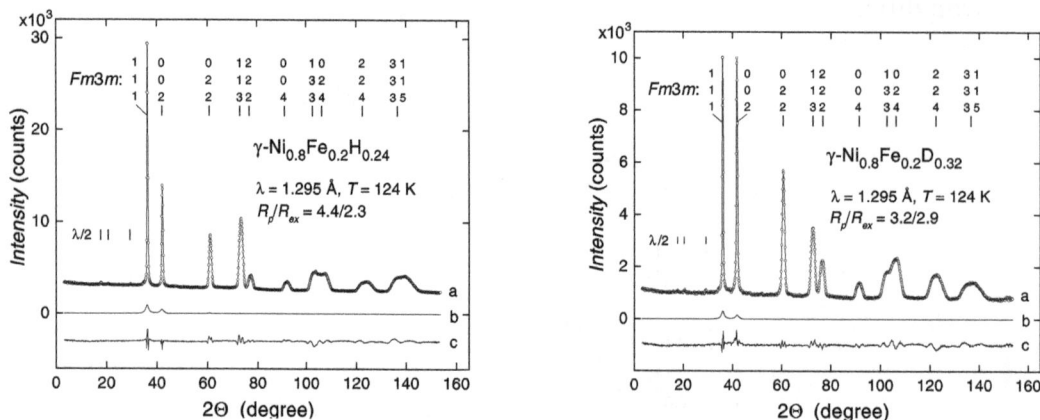

In figure above, Neutron diffraction patterns of the $Ni_{0.8}Fe_{0.2}H_{0.24}$ and $Ni_{0.8}Fe_{0.2}D_{0.32}$ powder samples measured at 124 K (circles) and results of their Rietveld analysis (solid curves). Curve a is the profile calculated for hydrogen/deuterium on octahedral interstices and with the magnetic contribution shown by curve b. Curve c is the difference between the experimental (circles) and calculated (curve a) spectra.

As seen from figure, the $Ni_{0.8}Fe_{0.2}H_{0.8}$ and $Ni_{0.8}Fe_{0.2}D_{0.63}$ samples were synthesized in the ranges of a steep increase in the x(P) isotherms. These are the ranges of supercritical anomalies of the isomorphous $\gamma_1 \leftrightarrow \gamma_2$ transformation terminating in a critical point at a lower temperature. A characteristic feature of such T −P ranges is a strong decrease in the H or D diffusion rate. In our case, the centres of the supercritical regions correspond to x ≈ 0.5. This explains the large inhomogeneity of the $Ni_{0.8}Fe_{0.2}D_{0.63}$ sample with the composition close to this critical value, the better homogeneity of the $Ni_{0.8}Fe_{0.2}H_{0.8}$ sample and a rather uniform distribution of H or D across the width of the samples with x < 0.3 and x > 0.9, which follows from the coincidence of the x-ray and neutron diffraction data for these samples.

To demonstrate that the inhomogeneous state of the samples with intermediate values of x was due to the low rate of H or D diffusion, one sample, $Ni_{0.8}Fe_{0.2}H_{0.64}$, labelled 3 in figures, was prepared differently from others. Prior to the 24 h hydrogenation at 3.66 GPa and 325 °C, this sample was exposed to the same hydrogen pressure at 380 °C for 8 h in order to accelerate the hydrogen diffusion and to achieve a more uniform initial hydrogen distribution over the sample volume. As can be seen from figure, the surface and the bulk properties of this sample were, indeed, similar. Its $\Delta V_a(x)$ value lies significantly below the dependence plotted as a dashed line and falls onto the straight solid line with a slope of $\beta = 2.15$ Å³ per metal atom per H(D) atom. This straight line is likely to represent the $\Delta V_a(x)$ dependence for homogeneous $Ni_{0.8}Fe_{0.2}Hx$ and $Ni_{0.8}Fe_{0.2}Dx$ solid solutions, as all the experimental $\Delta V_a(x)$ points for such solutions are close to it.

It also seems worth mentioning that,although the minimum diffusion rate and therefore the maximum inhomogeneity of the $Ni_{0.8}Fe_{0.2}Hx$ and $Ni_{0.8}Fe_{0.2}Dx$ solutions should be expected for hydrogen contents around x = 0.5, the maximum difference between the mean hydrogen content of thick samples and their nearly equilibrium surface content should be observed at a higher hydrogen concentration. This is because the number of hydrogen atoms that must diffuse through the sample surface to achieve the equilibrium concentration increases with increasing concentration, whereas the time of hydrogenation is fixed. The observed maximum deviation of the x-ray non-equilibrium $\Delta V_a(x)$ values from the straight solid line at x ≈ 0.7–0.8 agrees with this picture.

Thus, the $\Delta V_a(x)$ dependence for homogeneous $Ni_{0.8}Fe_{0.2}Hx$ and $Ni_{0.8}Fe_{0.2}D_x$ solid solutions is close to the straight line with a slope of $\beta = (\partial/\partial x)\,\Delta V_a(x) = 2.15$ Å³ per metal atom per H(D) atom. The non-linearity of the $\Delta V_a(x)$ dependences of the $Ni_{0.8}Fe_{0.2}Hx$ solid solutions constructed earlier and in the present paper was a result of two factors:

i. The samples of intermediate concentrations were prepared in the T–P range of supercritical anomalies of the isomorphous $\gamma_1 \leftrightarrow \gamma_2$ transformation, and they remained in an inhomogeneous, non-equilibrium state due to the low H diffusion rate, and

ii. The lattice parameters of these samples were determined by x-ray diffraction that scanned a thin surface layer where the hydrogen concentration was higher than in the bulk, and these parameters were then related to the mean hydrogen content of the samples.

Similar non-linear $\Delta V_a(x)$ dependences for hydrogen solid solutions in other fcc nickel alloys and in fcc palladium alloys are likely to have the same origin, because samples of intermediate compositions in these systems also were synthesized in the T –P ranges of supercritical anomalies and studied by x-ray diffraction.

[57]Fe Mossbauer Investigation

The local distribution of hydrogen around iron atoms can be examined by [57]Fe Mossbauer spectroscopy. It has been shown for a wide range of metal ¨ hydrides that the Mossbauer isomer shift S is proportional to the number of hydrogen atoms surrounding the iron probes, the proportionality constant ranging from 0.08 to 0.10 mm s⁻¹ per hydrogen atom. In Mossbauer transmission experiments, the measured isomer ¨ shift represents an average over the bulk of the investigated sample.

The samples used for Mossbauer spectroscopy were 0.08 mm thick foils hydrogenated at 250 °C. They were measured at 4.2 K with a source of ^{57}Co in rhodium also at a temperature of 4.2 K. The isomer shifts S of the studied samples are listed in table. The S(x)-dependence obtained is significantly non-linear at low x-values. This directly corresponds to a non-linear dependence of the number of hydrogen atoms around the Fe probes.

Table: The Mossbauer isomer shifts S for $Ni_{0.8}Fe_{0.2}Hx$ samples. NH is the estimated average number of H atoms surrounding an Fe atom; N_{Fe} is the average number of Fe atoms surrounding a H atom; $\langle N_{Fe} \rangle = 1.2$ is the average number of Fe atoms surrounding an arbitrary octahedral interstitial position.

x	S (mm s^{-1})	N_H	N_{Fe}	NFe/$\langle N_{Fe} \rangle$
0.000	−0.101	0.00	−	
0.044	−0.098	0.03	0.15	0.13
0.152	−0.080	0.24	0.32	0.27
0.829	0.310	4.84	1.17	0.98
0.846	0.338	5.17	1.22	1.02
0.973	0.399	5.89	1.21	1.01
0.995	0.406	5.97	1.20	1.00

Somewhat more quantitative information can be obtained by employing a rough model: assuming that in the nearly fully hydrogenated sample with x = 0.995 the remaining empty octahedral interstitials are randomly distributed over the sample, one obtains $N_H = 6 \times 0.995 = 5.97$ for the average number of hydrogen atoms around an iron (six being the number of octahedral interstitial positions surrounding an fcc metal site). Comparing the isomer shift S = 0.406 mm s^{-1} of this sample with $S_0 = -0.101$ mm s^{-1} for a hydrogen-free $Ni_{0.8}Fe_{0.2}$ sample yields dS = 0.085 mm s^{-1} for the isomer shift induced by a single hydrogen neighbour. With this value it is possible to calculate the mean number of hydrogen atoms around iron for all investigated samples. The results are shown in table.

It is instructive to also calculate the corresponding number N_{Fe} of iron atoms around a given hydrogen atom. By counting the number of Fe–H 'bonds' one arrives at $N_{Fe} = y\, NH/x$, where y is the iron content in a $Ni_{1-y}Fe_y$ alloy. N_{Fe} should be compared with the average number $\langle N_{Fe} \rangle = 6y = 1.2$ of Fe atoms surrounding an arbitrary octahedral interstitial position in a random alloy. As one can see, N_{Fe} is significantly smaller than $\langle N_{Fe} \rangle$ in the weakly loaded $Ni_{0.8}Fe_{0.2}$ samples, where the hydrogen selects sites with no or few iron neighbours. This shows that interstitial positions near Fe atoms are energetically less favourable for hydrogen than positions without or with relatively few Fe neighbours. The observed non-uniform distribution of hydrogen atoms over the interstitial positions is a typical example of partial thermodynamic equilibrium in a system where one component, namely hydrogen, can move freely, while the other components are frozen.

Iron(I) Hydride

Iron(I) hydride, systematically named iron hydride and poly(hydridoiron) is a solid inorganic compound with the chemical formula (FeH)$_n$ (also written ([FeH])$_n$ or FeH). It is both thermodynamically and kinetically unstable toward decomposition at ambient temperature, and as such, it little is known about its bulk properties.

Iron(I) hydride is the simplest polymeric iron hydride. Due to its instability, it has no practical industrial uses. However, in metallurgical chemistry, iron(I) hydride is fundamental to certain forms of iron-hydrogen alloys.

The systematic name *iron hydride*, a valid IUPAC name, is constructed according to the compositional nomenclature. However, as the name is compositional in nature, it does not distinguish between compounds of the same stoichiometry, such as molecular species, which exhibit distinct chemical properties. The systematic names *poly(hydridoiron)* and *poly[ferrane(1)]*, also valid IUPAC names, are constructed according to the additive and electron-deficient substitutive nomenclatures, respectively. They do distinguish the titular compound from the others.

Hydridoiron

Hydridoiron, also systematically named ferrane(1), is a related compound with the chemical formula FeH (also written [FeH]). It is also unstable at ambient temperature with the additional propensity to autopolymerize, and so cannot be concentrated.

Hydridoiron is the simplest molecular iron hydride. In addition, it may be considered to be the iron(I) hydride monomer. It has been detected in isolation only in extreme environments, like trapped in frozen noble gases, in the atmosphere of cool stars, or as a gas at temperatures above the boiling point of iron. It is assumed to have three dangling valence bonds, and is therefore a free radical; its formula may be written FeH^{3+} to emphasize this fact.

At very low temperatures (below 10 K), FeH may form a complex with molecular hydrogen $FeH \cdot H_2$.

Hydridoiron was first detected in the laboratory by B. Kleman and L. Åkerlind in the 1950s.

Properties

Radicality and Acidity

A single electron of another atomic or molecular species can join with the iron centre in hydridoiron by substitution:

$$[FeH] + RR \rightarrow [FeHR] + \cdot R$$

Because of this capture of a single electron, hydridoiron has radical character. Hydridoiron is a strong radical.

An electron pair of a Lewis base can join with the iron centre by adduction:

$$[FeH] + :L \rightarrow [FeHL]$$

Because of this capture of an adducted electron pair, hydridoiron has Lewis-acidic character. It should be expected that iron(I) hydride has significantly diminished radical properties, but has similar acid properties, however reaction rates and equilibrium constants are different.

Structure

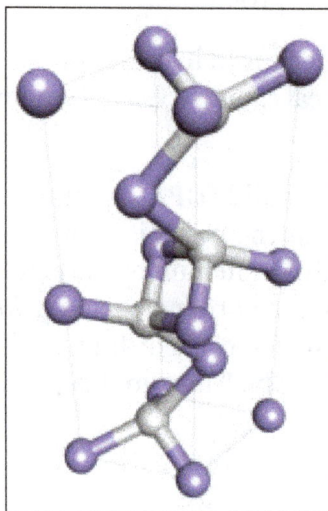

In iron(I) hydride, the atoms form a network, individual atoms being interconnected by covalent bonds. Since it is a polymeric solid, a monocrystalline sample is not expected to undergo state transitions, such as melting and dissolution, as this would require the rearrangement of molecular bonds and consequently, change its chemical identity. Colloidal crystalline samples, wherein intermolecular forces are relevant, are expected to undergo state transitions.

Iron(I) hydride adopts a double hexagonal close-packed crystalline structure with the $P6_3/mmc$ space group, also referred to as epsilon-prime iron hydride in the context of the iron-hydrogen system. It is predicted to exhibit polymorphism, transitioning at some temperature below −173 °C (−279 °F) to a face-centred crystalline structure with the Fm3m space group.

Electromagnetic Properties

band name	wavelength nm	wavenumber cm^{-1}	transition
Wing-Ford	989.6	10100	$F^4\Delta - X^4\Delta$
blue	490	20408	$g^6\Phi - a^6\Delta$
green	530	18867	$e^6\Pi - a^6\Delta$

FeH is predicted to have a quartet and a sextet ground states.

The FeH molecule has at least four low energy electronic states caused by the non bonding electron taking up positions in different orbitals: $X^4\Delta$, $a^6\Delta$ $b^6\Pi$, and $c^6\Sigma^+$. Higher energy states are termed $B^4\Sigma^-$, $C^4\Phi$, $D^4\Sigma^+$, $E^4\Pi$, and $F^4\Delta$. Even higher levels are labelled $G^4\Pi$ and $H^4\Delta$ from the quartet system, and $d^6\Sigma^-$, $e^6\Pi$, $f^6\Delta$, and $g^6\Phi$. In the quartet states the inner quantum number J takes on values 1/2, 3/2, 5/2, and 7/2.

FeH has an important absorption band (called the Wing-Ford band) in the near infrared with a band edge at 989.652 nm and a maximum absorption at 991 nm. It also has lines in the blue at 470 to 502.5 nm and in green from 520 to 540 nm.

The small isotope shift of the deuterated FeD compared to FeH at this wavelength shows that the band is due to a (0,0) transition from the ground state, namely $F^4\Delta - X^4\Delta$.

Various other bands exists in each part of the spectrum due to different vibrational transitions. The (1,0) band, also due to $F^4\Delta - X^4\Delta$ transitions, is around 869.0 nm and the (2,0) band around 781.8 nm.

Within each band there are a great number of lines. These are due to transition between different rotational states. The lines are grouped into subbands $^4\Delta_{7/2} - {}^4\Delta_{7/2}$ (strongest) and $^4\Delta_{5/2} - {}^4\Delta_{5/2}$, $^4\Delta_{3/2} - {}^4\Delta_{3/2}$ and $^4\Delta_{1/2} - {}^4\Delta_{1/2}$. The numbers like 7/2 are values for Ω the spin component. Each of these has two branches P and R, and some have a Q branch. Within each there is what is called Λ splitting that results in a lower energy lines (designated "a") and higher energy lines (called "b"). For each of these there is a series of spectral lines dependent on J, the rotational quantum number, starting from 3.5 and going up in steps of 1. How high J gets depends on the temperature. In addition there are 12 satellite branches $^4\Delta_{7/2} - {}^4\Delta_{5/2}$, $^4\Delta_{5/2} - {}^4\Delta_{3/2}$, $^4\Delta_{3/2} - {}^4\Delta_{1/2}$, $^4\Delta_{5/2} - {}^4\Delta_{7/2}$, $^4\Delta_{3/2} - {}^4\Delta_{5/2}$ and $^4\Delta_{1/2} - {}^4\Delta_{3/2}$ with P and R branches.

Some lines are magnetically sensitive, such as 994.813 and 995.825 nm. They are broadened by the Zeeman effect yet others in the same band are insensitive to magnetic fields like 994.911 and 995.677 nm. There are 222 lines in the (0-0) band spectrum.

Occurrence in Outer Space

Iron hydride is one of the few molecules found in the Sun. Lines for FeH in the blue-green part of the solar spectrum were reported in 1972, including many absorption lines in 1972. Also sunspot umbras show up the Wing-Ford band prominently.

Bands for FeH (and other hydrides of transition metals and alkaline earths) show up prominently in the emission spectra for M dwarfs and L dwarfs, the hottest kind of brown dwarf. For cooler T dwarfs, the bands for FeH do not appear, probably due to liquid iron clouds blocking the view of the atmosphere, and removing it from the gas phase of the atmosphere. For even cooler brown dwarfs (<1350 K), signals for FeH reappear, which is explained by the clouds having gaps.

The explanation for the kind of stars that the FeH Wing-Ford band appears in, is that the temperature is around 3000 K and pressure is sufficient to have a large number of FeH molecules formed. Once the temperature reaches 4000 K as in a K dwarf the line is weaker due to more of the molecules being dissociated. In M giant red giants the gas pressure is too low for FeH to form.

Elliptical and lenticular galaxies have also have an observable Wing-Ford band, due to a large amount of their light coming from M dwarfs.

Iron(II) Hydride

Iron(II) hydride, systematically named iron dihydride and poly(dihydridoiron) is solid inorganic compound with the chemical formula $(FeH_2)_n$ (also written $([FeH_2])_n$ or FeH_2).). It is kinetically unstable at ambient temperature, and as such, little is known about its bulk properties. However, it known as a black, amorphous powder, which was synthesised for the first time in 2014.

Iron(II) hydride is the second simplest polymeric iron hydride (after iron(I) hydride). Due to its instability, it has no practical industrial uses. However, in metallurgical chemistry, iron(II) hydride is fundamental to certain forms of iron-hydrogen alloys.

The systematic name *iron dihydride*, a valid IUPAC name, is constructed according to the compositional nomenclature. However, as the name is compositional in nature, it does not distinguish between compounds of the same stoichiometry, such as molecular species, which exhibit distinct chemical properties. The systematic names *poly(dihydridoiron)* and *poly[ferrane(2)]*, also valid IUPAC names, are constructed according to the additive and electron-deficient substitutive nomenclatures, respectively. They do distinguish the titular compound from the others.

Dihydridoiron

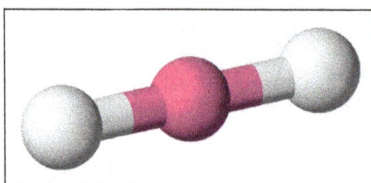

Dihydridoiron, also systematically named ferrane(2), is a related inorganic compound with the chemical formula FeH_2 (also written $[FeH_2]$). It is both kinetically unstable at concentration and at ambient temperature.

Dihydridoiron is the second simplest molecular iron hydride (after hydridoiron), and is also the progenitor of clusters with the same stoichiometry. In addition, it may be considered to be the iron(II) hydride monomer. It has been observed in matrix isolation.

Properties

Acidity and Basicity

An electron pair of a Lewis base can join with the iron centre in dihydridoiron by adduction:

$$[FeH_2] + \; :L \; \rightarrow [FeH_2L]$$

Because of this capture of an adducted electron pair, dihydridoiron has Lewis acidic character. Dihydridoiron has the capacity to capture up to four electron pairs from Lewis bases.

A proton can join with the iron centre by dissociative protonation:

$$FeH_2 + H^+ \rightarrow FeH^+ + H_2$$

Because dissociative protonation involves the capture of the proton (H^+) to form a Kubas complex ($[FeH(H_2)]^+$) as an intermediate, dihydridoiron and its adducts of weak-field Lewis bases, such as water, also have Brønsted–Lowry basic character. They have the capacity to capture up to two protons. Its dissociated conjugate acids are hydridoiron(1+) and iron(2+) (FeH^+ and Fe^{2+}).

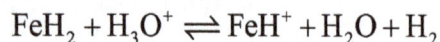

$$FeH_2 + H_3O^+ \rightleftharpoons FeH^+ + H_2O + H_2$$

Aqueous solutions of adducts of weak-field Lewis bases are however, unstable due to hydrolysis of the dihydridoiron and hydridoiron(1+) groups:

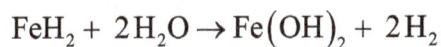

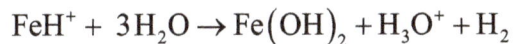

$$FeH_2 + 2H_2O \rightarrow Fe(OH)_2 + 2H_2$$

$$FeH^+ + 3H_2O \rightarrow Fe(OH)_2 + H_3O^+ + H_2$$

It should be expected that iron dihydride clusters and iron(II) hydride have similar acid-base properties, although reaction rates and equilibrium constants are different.

Alternatively, a hydrogen centre in the dihydridoiron group in adducts of strong-field Lewis bases, such as carbon monoxide, may separate from the molecule by ionisation:

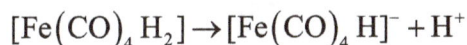

$$[Fe(CO)_4 H_2] \rightarrow [Fe(CO)_4 H]^- + H^+$$

Because of this release of the proton, adducts of strong-field Lewis bases have may have Brønsted–Lowry acidic character. They have the capacity to release up to two protons.

$$[Fe(CO)_4 H_2] + H_2O \rightleftharpoons [Fe(CO)_4 H]^- + H_3O^+$$

Mixed adducts with Lewis bases of differing fields strengths may exhibit intermediate behaviour.

Structure

In iron(II) hydride, the atoms form a network, individual atoms being interconnected by covalent bonds. Since it is a polymeric solid, a monocrystalline sample is not expected to undergo state transitions, such as melting and dissolution, as this would require the rearrangement of molecular bonds and consequently, change its chemical identity. Colloidal crystalline samples, wherein inter-molecular forces are relevant, are expected to undergo state transitions.

At least up to −173 °C (−279 °F), iron(II) hydride is predicted to have a body-centred tetragonal crystalline structure with the I4/mmm space group. In this structure, iron centres have a capped square-antiprismatic coordination geometry, and hydrogen centres have square-planar and square-pyramidal geometries.

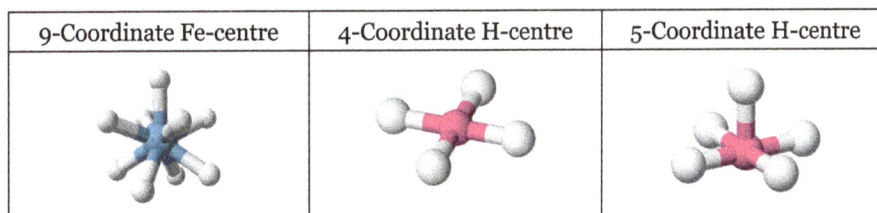

9-Coordinate Fe-centre	4-Coordinate H-centre	5-Coordinate H-centre

An amorphous form of iron(II) hydride is also known.

The infrared spectrum for dihydridoiron shows that the molecule has a linear H–Fe–H structure in the gas phase, with an equilibrium distance between the iron atom and the hydrogen atoms of 0.1665 nm.

Electronic Properties

A few of dihydridoiron's electronic states lie relatively close to each other, giving rise to varying degrees of radical chemistry. The ground state and the first two excited states are all quintet radicals with four unpaired electrons ($X^5\Delta_g$, $A^5\Pi_g$, $B^5\Sigma_g{}^+$). With the first two excited states only 22 and 32 kJ mol^{-1} above the ground state, a sample of dihydridoiron contains trace quantities of excited states even at room temperature. Furthermore, Crystal field theory predicts that the low transition energies correspond to a colourless compound.

The ground electronic state is $^5\Delta_g$.

State Transitions of ^{56}FeH$_2$ in the v_3 Fundamental Band

Transition	Wavenumber (cm^{-1})	Frequency (THz)
$P_4(10)$	1614.912	48.4100
$P_4(7)$	1633.519	48.9717
$Q_4(4),Q_3(3)$	1672.658	50.1450
$Q_4(4),Q_4(5),Q_3(3)$	1676.183	50.2507
$R_4(4)$	1704.131	51.0886
$R_4(5)$	1707.892	51.2013
$R_4(8)$	1725.227	51.7210
$R_4(9)$	1729.056	52.8358

Metallurgical Chemistry

In iron-hydrogen alloys that have hydrogen content near 3.48 wt%, hydrogen can precipitate as iron(II) hydride and lesser quantities of other polymeric iron hydrides. However, due to the limited solubility of hydrogen in iron, the optimum content for the formation of iron(II) hydride can only be reached by applying extreme pressure.

In metallurgical chemistry, iron(II) hydride is fundamental to certain forms of iron-hydrogen alloys. It occurs as a brittle component within the solid matrix, with a physical makeup that depends on its formation conditions and subsequent heat treatment. As it decomposes over time, the alloy will slowly become softer and more ductile, and may start to suffer from hydrogen embrittlement.

Production

Dihydridoiron has been produced by several means, including:

By reaction of FeCl$_2$ and PhMgBr under a hydrogen atmosphere (1929).

- Electrical discharge in a mixture of pentacarbonyliron and dihydrogen diluted in helium at 8.5 Torr.

- Evaporation of iron with a laser in an atmosphere of hydrogen, pure or diluted in neon or argon, and condensing the products on a cold surface below 10 K.

- Decomposition product of collision-excited ferrocenium ions.

$$nFe + nH_2 \rightarrow (FeH_2)_n$$

The process involves iron(I) hydride as an intermediate, and occurs in two steps.

1. $2nFe + nH_2 \rightarrow 2(FeH)_n$
2. $2(FeH)_n + nH_2 \rightarrow 2(FeH_2)_n$

Bis[bis(mesityl)iron] Reduction

Amorphous iron(II) hydride is produced by bis[bis(mesityl)iron] reduction. In this process, bis-bis(mesityl)iron] is reduced with hydrogen under an applied pressure of 100 atmospheres to produce iron(II) hydride according to the reaction:

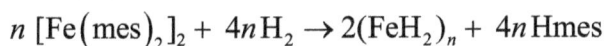

$$n\,[Fe(mes)_2]_2 + 4n\,H_2 \rightarrow 2(FeH_2)_n + 4n\,Hmes$$

The process involves bis[hydrido(mesityl)iron] and dihydridoiron as intermediates, and occurs in three steps.

1. $[Fe(mes)_2]_2 + 2H_2 \rightarrow [FeH(mes)]_2 + 2\,Hmes$
2. $[FeH(mes)]_2 + H_2 \rightarrow FeH_2 + Hmes$
3. $n\,FeH_2 \rightarrow (FeH_2)_n$

Reactions

As dihydridoiron is an electron-deficient molecule, it spontaneously autopolymerises in its pure form, or converts to an adduct upon treatment with a Lewis base. Upon treatment of adducts of weak-field Lewis bases with a dilute standard acid, it converts to an hydridoiron(1+) salt and elemental hydrogen. Treatment of adducts of strong-field Lewis bases with a standard base, converts it to a metal ferrate(1−) salt and water. Oxidation of iron dihydrides give iron(II) hydroxide, whereas reduction gives hexahydridoferrate(4−) salts. Unless cooled to at most−243 °C (−405.4 °F), dihydridoiron decomposes to produce elemental iron and hydrogen. Other iron dihydrides and adducts of dihydridoiron decompose at higher temperatures to also produce elemental hydrogen, and iron or polynuclear iron adducts:

$$FeH_2 \rightarrow Fe + H_2$$

Non-metals, including oxygen, strongly attack iron dihydrides, forming hydrogenated compounds and iron(II) compounds:

$$FeH_2 + O_2 \rightarrow FeO + H_2O$$

Iron(II) compounds can also be prepared from an iron dihydride and an appropriate, concentrated acid:

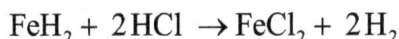

$$FeH_2 + 2HCl \rightarrow FeCl_2 + 2H_2$$

Ferroalloy

Ferroallo is an alloy of iron (less than 50 percent) and one or more other metals, important as a source of various metallic elements in the production of alloy steels. The principal ferroalloys are ferromanganese, ferrochromium, ferromolybdenum, ferrotitanium, ferrovanadium, ferrosilicon, ferroboron, and ferrophosphorus. These are brittle and unsuitable for direct use in fabricating products, but they are useful sources of these elements for the alloy steels. Ferroalloys usually have lower melting ranges than the pure elements and can be incorporated more readily in the molten steel. They are added to liquid steel to achieve a specified chemical composition and provide properties needed to make particular products. They are in fact used in all steels—e.g., plain carbon, stainless, alloy, electrical, tool, and so on.

Ferroalloys are prepared from charges of the nonferrous metal ore, iron or iron ore, coke or coal, and flux by treatment at high temperature in submerged-arc electric furnaces. An aluminothermic reduction process is used for making ferrovanadium, ferrotitanium, and ferroniobium (ferrocolumbium).

Various ferroalloys are used in the steel-making process to improve the performance of steel as industrial materials. Traditionally, additive alloys were developed in the form of iron alloys which became known as ferroalloys due to their high iron content. In response to the recent progress in steel-making technology, the class of ferroalloys has expanded to include additive alloys without iron content. Still, ferroalloys of chromium, manganese, and silicon are consumed in far greater tonnage in steel-making than any other additive alloys.

Ferroalloys are produced in several types of furnaces, depending on the chemistry as well as the thermal and electrical parameters of the specific process. The most used are submerged arc furnaces (SAFs) and resistance-heating furnaces (RHFs), where heating of the charge, slag, and alloy proceeds by the application of electric current. In these electric furnaces, the tips of electrodes protrude into the furnace hearth and become hidden by a burden (charge) of raw materials, which have been fed into the furnace from the top. The burden typically is partially transformed by chemical reactions appropriate for the respective reaction zones in the furnace at the local temperature and pressure. The main difference between SAFs and RHFs is whether or not an electric arc is present at the tip or along the side of the electrode. For example, in silicon and ferrosilicon production, the temperature required for the process to run with an acceptable silicon yield is above 1900°C. This requires an intense heat source and therefore dictates the formation of an electric arc, rather than heating by ohmic resistance (joule heating). The latter proceeds when electric current is conducted through the charge or slag surrounding the electrodes. Several ferroalloys smelting processes do not require an arc in order to reach the necessary thermodynamic conditions, but in many cases there are strong indications that an arc may nevertheless be present, such as in ferrochromium or ferromanganese production.

As the sole purpose of the electric current passing through the SAF is to release heat to fuel the chemical processes in the furnace, the current can be either AC or DC. The most common configuration for an industrial scale furnace is three-phase AC, where three electrodes are embedded into the furnace raw material charge. Three AC-current phases separated by a phase shift of 120° pass through the respective electrodes and cancel out at a star point located in the center of the furnace. One phase AC is used mostly in smaller laboratory or pilot scale furnaces, and it requires a top and a bottom electrode. DC furnaces also require a bottom electrode in addition to the consumable electrode from the top. Both small scale and large industrial scale DC furnaces are in operation around the world.

Ferroalloy Production

The ferroalloy industry is associated with the iron and steel industries, its largest customers. Ferroalloys impart distinctive qualities to steel and cast iron and serve important functions during iron and steel production cycles. The principal ferroalloys are those of chromium, manganese, and silicon. Chromium provides corrosion resistance to stainless steels. Manganese is essential to counteract the harmful effects of sulfur in the production of virtually all steels and cast iron. Silicon is used primarily for deoxidation in steel and as an alloying agent in cast iron. Boron, cobalt, columbium, copper, molybdenum, nickel, phosphorus, titanium, tungsten, vanadium, zirconium, and the rare earths impart specific characteristics and are usually added as ferroalloys.

Process Description

A typical ferroalloy plant is illustrated in figure. A variety of furnace types, including submerged electric arc furnaces, exothermic (metallothermic) reaction furnaces, and electrolytic cells can be used to produce ferroalloys.

Submerged Electric Arc Process

In most cases, the submerged electric arc furnace produces the desired product directly. It may produce an intermediate product that is subsequently used in additional processing methods. The submerged arc process is a reduction smelting operation. The reactants consist of metallic ores (ferrous oxides, silicon oxides, manganese oxides, chrome oxides, etc.) and a carbon-source reducing agent, usually in the form of coke, charcoal, high- and low-volatility coal, or wood chips. Limestone may also be added as a flux material. Raw materials are crushed, sized, and, in some cases, dried, and then conveyed to a mix house for weighing and blending. Conveyors, buckets, skip hoists, or cars transport the processed material to hoppers above the furnace. The mix is then gravity-fed through a feed chute either continuously or intermittently, as needed. At high temperatures in the reaction zone, the carbon source reacts with metal oxides to form carbon monoxide and to reduce the ores to base metal. A typical reaction producing ferrosilicon is shown below:

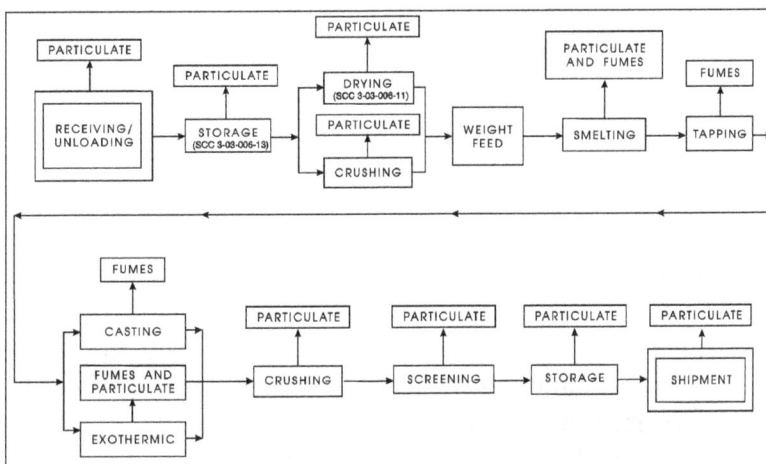

$$Fe_2O_3\ 2SiO_2 + 7C \rightarrow 2\,FeSi + 7\,CO$$

Typical ferroalloy production process.

Table: Ferroalloy Processes and Respective Product Groups.

Process	Product
Submerged arc furnace	Silvery iron (15-22% Si)
	Ferrosilicon (50% Si)
	Ferrosilicon (65-75% Si)
	Silicon metal
	Silicon/manganese/zirconium (SMZ)
	High carbon (HC) ferromanganese
	Siliconmanganese
	HC ferrochrome
	Ferrochrome/silicon
	FeSi (90% Si)
Exothermic Silicon reduction	Low carbon (LC) ferrochrome, LC ferromanganese, medium carbon (MC) ferromanganese
Aluminum Reduction	Chromium metal, ferrotitanium, ferrocolumbium, ferovanadium
Mixed aluminothermal/silicothermal	Ferromolybdenum, ferrotungsten
Electrolytic	Chromium metal, manganese metal
Vacuum furnace	LC ferrochrome
Induction furnace	Ferrotitanium

Smelting in an electric arc furnace is accomplished by conversion of electrical energy to heat. An alternating current applied to the electrodes causes current to flow through the charge between the electrode tips. This provides a reaction zone at temperatures up to 2000°C (3632°F). The tip of each electrode changes polarity continuously as the alternating current flows between the tips. To maintain a uniform electric load, electrode depth is continuously varied automatically by mechanical or hydraulic means.

A typical submerged electric arc furnace design is depicted in figure. The lower part of the submerged electric arc furnace is composed of a cylindrical steel shell with a flat bottom or hearth. The interior of the shell is lined with 2 or more layers of carbon blocks. The furnace shell may be water-cooled to protect it from the heat of the process. A water-cooled cover and fume collection hood are mounted over the furnace shell. Normally, 3 carbon electrodes arranged in a triangular formation extend through the cover and into the furnace shell opening. Prebaked or selfbaking (Soderberg) electrodes ranging from 76 to over 100 cm (30 to over 40 inches) in diameter are typically used. Raw materials are sometimes charged to the furnace through feed chutes from above the furnace. The surface of the furnace charge, which contains both molten material and unconverted charge during operation, is typically maintained near the top of the furnace shell. The lower ends of the electrodes are maintained at about 0.9 to 1.5 meters (3 to 5 feet) below the charge surface. Threephase electric current arcs from electrode to electrode, passing through the charge material. The charge material melts and reacts to form the desired product as the electric energy is converted into heat. The carbonaceous material in the furnace charge reacts with oxygen in the metal oxides of the charge and reduces them to base metals. The reactions produce large quantities of carbon monoxide (CO) that passes upward through the furnace charge. The molten metal and slag are removed (tapped) through 1 or more tap holes extending through the furnace shell at the hearth level. Feed materials may be charged continuously or intermittently. Power is applied continuously. Tapping can be intermittent or continuous based on production rate of the furnace.

Submerged electric arc furnaces are of 2 basic types, open and covered. Most of the submerged electric arc furnaces in the U. S. are open furnaces. Open furnaces have a fume collection hood at least 1 meter (3.3 feet) above the top of the furnace shell. Moveable panels or screens are sometimes used to reduce the open area between the furnace and hood, and to improve emissions capture efficiency. Carbon monoxide rising through the furnace charge burns in the area between the charge surface and the capture hood. This substantially increases the volume of gas the containment system must handle. Additionally, the vigorous open combustion process entrains finer material in the charge. Fabric filters are typically used to control emissions from open furnaces.

Covered furnaces may have a water-cooled steel cover that fits closely to the furnace shell. The objective of covered furnaces is to reduce air infiltration into the furnace gases, which reduces combustion of that gas. This reduces the volume of gas requiring collection and treatment. The cover has holes for the charge and electrodes to pass through. Covered furnaces that partially close these hood openings with charge material are referred to as "mix-sealed" or "semi-enclosed furnaces". Although these covered furnaces significantly reduce air infiltration, some combustion still occurs under the furnace cover. Covered furnaces that have mechanical seals around the electrodes and sealing compounds around the outer edges are referred to as "sealed" or "totally closed". These furnaces have little, if any, air infiltration and undercover combustion. Water leaks from the cover into the furnace must be minimized as this leads to excessive gas production and unstable furnace operation. Products prone to highly variable releases of process gases are typically not made in covered furnaces for safety reasons. As the degree of enclosure increases, less gas is produced for capture by the hood system and the concentration of carbon monoxide in the furnace gas increases. Wet scrubbers are used to control emissions from covered furnaces. The scrubbed, high carbon monoxide content gas may be used within the plant or flared.

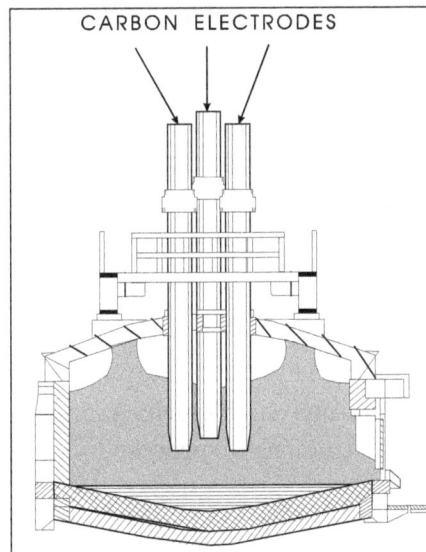

Typical submerged arc furnace design.

The molten alloy and slag that accumulate on the furnace hearth are removed at 1 to 5-hour intervals through the tap hole. Tapping typically lasts 10 to 15 minutes. Tap holes are opened with pellet shot from a gun, by drilling, or by oxygen lancing. The molten metal and slag flow from the tap hole into a carbon-lined trough, then into a carbon-lined runner that directs the metal and slag into a reaction ladle, ingot molds, or chills. (Chills are low, flat iron or steel pans that provide rapid cooling of the molten metal.) After tapping is completed, the furnace is resealed by inserting a carbon paste plug into the tap hole.

Chemistry adjustments may be necessary after furnace smelting to achieve a specified product. Ladle treatment reactions are batch processes and may include metal and alloy additions.

During tapping, and/or in the reaction ladle, slag is skimmed from the surface of the molten metal. It can be disposed of in landfills, sold as road ballast, or used as a raw material in a furnace or reaction ladle to produce a chemically related ferroalloy product.

After cooling and solidifying, the large ferroalloy castings may be broken with drop weights or hammers. The broken ferroalloy pieces are then crushed, screened (sized), and stored in bins until shipment. In some instances, the alloys are stored in lump form in inventories prior to sizing for shipping.

Exothermic (Metallothermic) Process

The exothermic process is generally used to produce high-grade alloys with low-carbon content. The intermediate molten alloy used in the process may come directly from a submerged electric arc furnace or from another type of heating device. Silicon or aluminum combines with oxygen in the molten alloy, resulting in a sharp temperature rise and strong agitation of the molten bath. Low- and medium-carbon content ferrochromium (FeCr) and ferromanganese (FeMn) are produced by silicon reduction. Aluminum reduction is used to produce chromium metal, ferrotitanium, ferrovanadium, and ferrocolumbium. Mixed alumino/silico thermal processing is used for producing ferromolybdenum and ferrotungsten. Although aluminum is more expensive than carbon or silicon, the products are purer. Low-carbon (LC) ferrochromium is typically produced by fusing chromium ore and lime in a furnace. A specified amount is then placed in a ladle (ladle No. 1). A known amount of an intermediate grade ferrochromesilicon is then added to the ladle. The reaction is extremely exothermic and liberates chromium from its ore, producing LC ferrochromium and a calcium silicate slag. This slag, which still contains recoverable chromium oxide, is reacted in a second ladle (ladle No. 2) with molten high-carbon ferrochromesilicon to produce the intermediategrade ferrochromesilicon. Exothermic processes are generally carried out in open vessels and may have emissions similar to the submerged arc process for short periods while the reduction is occurring.

Electrolytic Processes

Electrolytic processes are used to produce high-purity manganese and chromium. As of 1989, there were 2 ferroalloy facilities using electrolytic processes.

Manganese may be produced by the electrolysis of an electrolyte extracted from manganese ore or manganese-bearing ferroalloy slag. Manganese ores contain close to 50 percent manganese; furnace slag normally contains about 10 percent manganese. The process has 5 steps: (1) roasting the ore to convert it to manganese oxide (MnO), (2) leaching the roasted ore with sulfuric acid (H2SO4) to solubilize manganese, (3) neutralization and filtration to remove iron and aluminum hydroxides, (4) purifying the leach liquor by treatment with sulfide and filtration to remove a wide variety of metals, and (5) electrolysis.

Electrolytic chromium is generally produced from high-carbon ferrochromium. A large volume of hydrogen gas is produced by dissolving the alloy in sulfuric acid. The leachate is treated with ammonium sulfate and conditioned to remove ferrous ammonium sulfate and produce a chromealum for feed to the electrolysis cells. The electrolysis cells are well ventilated to reduce ambient hydrogen and hexavalent chromium concentrations in the cell rooms.

Emissions and Controls

Particulate is generated from several activities during ferroalloy production, including raw material handling, smelting, tapping, and product handling. Organic materials are generated almost exclusively from the smelting operation.

Particulate emissions from electric arc furnaces in the form of fumes account for an estimated 94 percent of the total particulate emissions in the ferroalloy industry. Large amounts of carbon monoxide and organic materials also are emitted by submerged electric arc furnaces. Carbon monoxide is formed as a byproduct of the chemical reaction between oxygen in the metal oxides of the charge and carbon contained in the reducing agent (coke, coal, etc.). Reduction gases containing organic compounds and carbon monoxide continuously rise from the high-temperature reaction zone, entraining fine particles and fume precursors. The mass weight of carbon monoxide produced sometimes exceeds that of the metallic product. The heat-induced fume consists of oxides of the products being produced and carbon from the reducing agent. The fume is enriched by silicon dioxide, calcium oxide, and magnesium oxide, if present in the charge.

In an open electric arc furnace, virtually all carbon monoxide and much of the organic matter burns with induced air at the furnace top. The remaining fume, captured by hooding about 1 meter above the furnace, is directed to a gas cleaning device. Fabric filters are used to control emissions from 85 percent of the open furnaces in the U. S. Scrubbers are used on 13 percent of the furnaces, and electrostatic precipitators on 2 percent.

Two emission capture systems, not usually connected to the same gas cleaning device, are necessary for covered furnaces. A primary capture system withdraws gases from beneath the furnace cover. A secondary system captures fumes released around the electrode seals and during tapping. Scrubbers are used almost exclusively to control exhaust gases from sealed furnaces. The scrubbers capture a substantial percentage of the organic emissions, which are much greater for covered furnaces than open furnaces. The gas from sealed and mix-sealed furnaces is usually flared at the exhaust of the scrubber. The carbon monoxide-rich gas is sometimes used as a fuel in kilns and sintering machines. The efficiency of flares for the control of carbon monoxide and the reduction of VOCs has been estimated to be greater than 98 percent. A gas heating reduction of organic and carbon monoxide emissions is 98 percent efficient.

Tapping operations also generate fumes. Tapping is intermittent and is usually conducted during 10 to 20 percent of the furnace operating time. Some fumes originate from the carbon lip liner, but most are a result of induced heat transfer from the molten metal or slag as it contacts the runners, ladles, casting beds, and ambient air. Some plants capture these emissions to varying degrees with a main canopy hood. Other plants employ separate tapping hoods ducted to either the furnace emission control device or a separate control device. Emission factors for tapping emissions are unavailable due to lack of data.

After furnace tapping is completed, a reaction ladle may be used to adjust the metallurgy by chlorination, oxidation, gas mixing, and slag metal reactions. Ladle reactions are an intermittent process, and emissions have not been quantified. Reaction ladle emissions are often captured by the tapping emissions control system.

Production by Processes

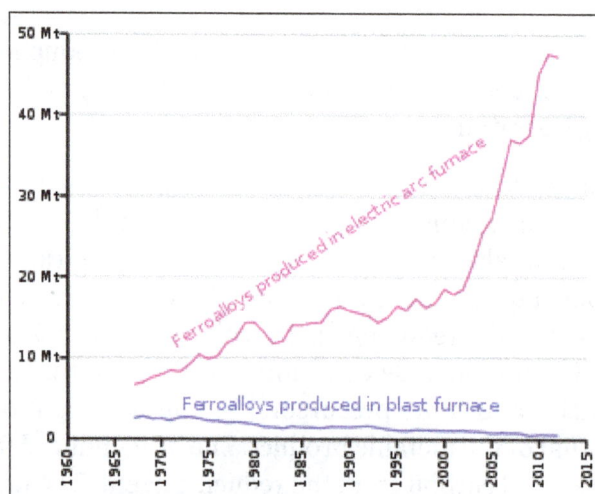

Evolution of the global ferroalloys production, by processes.

Ferroalloys are produced generally by two methods : in a blast furnace or in an electric arc furnace. Blast furnace production continuously decreased during the 20th century, whereas the electric arc production is still increasing. Today, feromanganese can be still efficiently produced in a blast furnace, but, even in this case, electric arc furnace are spreading. More commonly, ferroalloys are produced by carbothermic reactions, involving reduction of oxides with carbon (as coke) in the presence of iron. Some ferroalloys are produced by the addition of elements into molten iron.

It is also possible to produce somme ferroalloys by direct reduction processes. For example, the Krupp-Renn process is used in Japan to produce ferronickel.

Production and Consumption by Ferroalloys

Ferrochromium

The leading world chromite ore-producing countries in 2008 were India (almost 4 Mt), Kazakhstan (more than 3 Mt), and South Africa (almost 10 Mt). More than 94% of chromite ore production was smelted in electric-arc furnaces to produce ferrochromium for the metallurgical industry. The leading world ferrochromium-producing countries in 2008 were China (more than 1 Mt),

Kazakhstan (more than 1 Mt), and South Africa (more than 3 Mt). India and Russia each produced in excess of 0.5 Mt of ferrochromium. Most of the 7.84 Mt of ferrochromium produced worldwide was consumed in the manufacture of stainless steel which exceeded 26 Mt in 2008. Production for this is very economical.

Ferrochromium

Ferromanganese

Two manganese ferroalloys, ferromanganese and silicomanganese, are key ingredients for steelmaking. China is the leading world producer of manganese ferroalloys (2.7 Mt), with output about much greater than that of the next three major producers—Brazil (0.34 Mt), South Africa (0.61 Mt) and Ukraine (0.38 Mt)—combined.

Ferromolybdenum

Major producers of ferromolybdenum are Chile (16,918 t), China (40,000 t) and the United States which accounted for about 78% of world production of molybdenite ore in 2008, whereas Canada, Mexico and Peru accounted for the remainder. Molybdenite concentrates are roasted to form molybdic oxide, which can be converted into ferromolybdenum, molybdenum chemicals, or molybdenum metal. Although the United States was the second leading molybdenum-producing country in the world in 2008, it imported more than 70% of its ferromolybdenum requirements in 2008, mostly for the steel industry (83% of ferromolybdenum consumed).

Ferronickel

In 2008, the major ferronickel-producing countries were Japan (301,000 t), New Caledonia (144,000 t) and Colombia (105,000 t). Together, these three countries accounted for about 51% of world production if China is excluded. Ukraine, Indonesia, Greece, and Macedonia, in descending order of gross weight output, all produced between 68,000 t and 90,000 t of ferronickel, accounting for an additional 31%, excluding China. China was excluded from statistics because its industry produced large tonnages of nickel pig iron in addition to a spectrum of conventional ferronickel grades, for an estimated combined output of 590,000 t gross weight. The nickel content of individual Chinese products varied from about 1.6% to as much as 80%, depending upon customer end use.

In the United States, the steel industry accounted for virtually all the ferronickel consumed in 2008, with more than 98% used in stainless and heat-resistant steels; no ferronickel was produced in the US in 2008.

The nickel pig iron is a low grade ferronickel made in China, which is very popular since the 2010s.

Ferrosilicon

Ferrosilicon

Silicon ferroalloy consumption is driven by cast iron and steel production, where silicon alloys are used as deoxidizers. Some silicon metal was also used as an alloying agent with iron. On the basis of silicon content, net production of ferrosilicon and miscellaneous silicon alloys in the US was 148,000 t in 2008. China is the major supplier, which in 2008 produced more ferrosilicon (4.9 Mt) than the rest of the world combined. Other major manufacturers are Norway (0.21 Mt), Russia (0.85 Mt) and US (0.23 Mt).

Ferrotungsten

Tungsten is an important alloying element in high-speed and other tool steels, and is used to a lesser extent in some stainless and structural steels. Tungsten is often added to steel melts as ferrotungsten, which can contain up to 80% tungsten. World ferrotungsten production is dominated by China, which in 2008 exported 4,835 t (gross weight) of the alloy. Ferrotungsten is relatively expensive, with the prices around $31–44 per kilogram of contained tungsten.

Ferrotitanium

Ferrotitanium is the alloying addition consisting of iron and titanium with the minimum titanium weight content of 20 % and the maximum weight content of 75 % manufactured by reduction or melting. Manufacture of titanium alloy ingots and mill products also results in generation of metal scrap. Some scrap is graded substandard as it cannot be recycled to charge materials for melting titanium alloy ingots due to oxidation degree, pinches and forge laps. In order to use substandard titanium scrap, the manufacturing process for high-grade ferrotitanium has been developed and mastered. It is based on melting ingots of low-melting eutectic (1085°) in the titanium-iron system with 65-75 % of titanium. This process allows to melt scrap in the form of solids, bundle and briquetted clippings and loose crushed titanium chips. Obsolete titanium scrap has become more used lately. In order to manufacture ferrotitanium with the specified amount of additions, all titanium scrap is sorted based on alloys and alloy groups.

High-grade ferrotitanium is manufactured by VSMPO-AVISMA Corporation by induction and electro-slag remelting. Induction ferrotitanium is produced in channel-type induction furnaces in the form of a flat taper ingot with the weight of 500-550 kg. Each heat is subject to chemical analysis. Sampling is carried out from liquid metal pool prior to pouring the melt into the mold. Titanium is an active element and takes nitrogen and oxygen as a result of air interaction. That is why it is required to protect the melt from gas saturating.

Electro-slag remelting (ESR) is one of the processes to manufacture quality ferrotitanium. High-grade ferrotitanium is produced by electro-slag remelting in electric furnaces with a copper water-cooled crucible. ESR furnaces use direct alternation current with controlled frequency. Such furnaces are characterized by low inductance, high efficiency factor and uniform heat flow. Charge materials are composed of chips of unalloyed steels and titanium alloys.

Induction furnace melting.

Ferro Titanium is used by stainless steel makers as a stabilizer to prevent chromium carbide forming at grain boundaries and in the production of low carbon steels for sheet production. Ferro Titanium is also used in foundries for addition of titanium to the molten metal imparting excellent strength to it without altering other ratios.

Various titanium bearing materials are employed in making Ferro Titanium. Major sources include titanium scrap, titanium sponge, ilmenite, leucoxene, rutile. These materials are either reduced by aluminothermic means or melted in induction furnace with iron to produce ferro titanium of various grades and specifications. Various grades are manufactured, the most common being 20-25% Ti for foundries, 35-40% Ti for welding, 70% Ti for stainless steel production.

Table: Chemical Specifications of various grades of Ferro Titanium.

Element	MAC-FeTi-20	MAC-FeTi-30	MAC-FeTi-35	MAC-FeTi-70
Ti	20 % min	30 % min	35 % min	70 % min
Al	8 % max	1 % max	6 % max	1 % max
C	0.20 % max	0.10 % max	0.10 % max	0.10 % max
Si	3 % max	3 % max	3 % max	3 % max
S	0.10 % max	0.10 % max	0.10 % max	0.05 % max
P	0.10 % max	0.10 % max	0.10 % max	0.05 % max

Ferroaluminum

Ferro Aluminium is a ferro alloy comprising of Iron and Aluminium in a ratio of approximately 35% to 65%.

It is used as a deoxidising agent in special steels. When a low melting point is required, more ferro aluminium is added.

The manufacturing of Ferro Aluminum is performed as a three step process:

- Initially, Alumina (Al_2O_3) is obtained through the Bayer process by digestion of bauxite with sodium hydroxide (NaOH) at about 240 °C.

- Consequently, the alumina is subjected to a Hall electrolytic process together with Cryolite to obtain Aluminium that will then be combined with Iron to obtain the Ferro Aluminium.

- Finally, after the solidification of the metal, milling and sieving processes are carried out, thus obtaining the suitable particle size for its addition in steel and cast iron.

Most of the world's Ferro Aluminium supply is produced by Australia, China, Russia, USA and Canada, with the cost of electricity being the decisive factor in the aluminium obtaining process.

Uses

The uses of Ferro Aluminium include:

- Its ability to be a deoxidizing agent in the manufacturing of steels.

- The possibility to manufacture low melting point alloys.

- Its ability to carry out aluminothermic welding.

Ferrovanadium

Ferrovanadium is a master alloy with a vanadium content of at least 50% produced by reduction of the raw materials with carbon or aluminum. The high affinity of vanadium for carbon leads to carbide formation, so that carbothermic reduction can only be used when there is no requirement for vanadium with low carbon content. The reaction of V_2O_5 with aluminum is self-sustaining and the combustion time is only a few minutes. After cooling, the furnace is dismantled, and the block of metal is cleaned, then crushed to the desired particle size.

Ferrovanadium is produced by the reduction of V_2O_5 by silicon from ferrosilicon in the presence of lime in the electric arc furnace (3 MVA). The technology used for the smelting of ferrovanadium in the electric arc furnace utilizes three periods. The purpose of the first period is to restore vanadium from the current (return) fusion products of the third period of the previous heat. The charge, consisting of a recycled furnace slag, lime flux, ferrosilicon, aluminum, vanadium pentoxide, and iron chips, is loaded. After the first period (80 to 100 min), V_2O_5 content in the slag is reduced to 0.25% to 0.35%, and the metal ends with 25% to 30% V, 20% to 24% Si, and 0.3% to 0.5% C. The slag with this low content of vanadium is tapped from the furnace.

Composition

Vanadium content in ferrovanadium ranges from 35% to 85%. FeV80 (80% Vanadium) is the most common ferrovanadium composition. In addition to iron and vanadium, small amounts of silicon, aluminum, carbon, sulfur, phosphorus, arsenic, copper, and manganese are found in ferrovanadium. Impurities can make up to 11% by weight of the alloy. Concentrations of these impurities determine the grade of ferrovanadium.

Elemental Composition (Maximum % Weight)										
		V	Si	Al	C	S	P	As	Cu	Mn
Grade of Ferrovanadium	FeV75C0.1	70-85	0.8	2.0	0.1	0.05	0.05	0.05	0.1	0.4
	FeV75C0.15	70-85	1.0	2.5	0.15	0.1	0.1	0.05	0.1	0.6
	FeV50C0.4	48-60	1.8	0.2	0.4	0.02	0.07	0.01	0.2	2.7
	FeV50C0.5	48-60	2.0	0.3	0.5	0.02	0.07	0.01	0.2	4.0
	FeV50C0.6	48-60	2.0	0.3	0.6	0.03	0.07	0.02	0.2	5.0
	FeV50C0.3	> 50	2.0	2.5	0.3	0.1	0.1	0.05	0.2	0.2
	FeV50C0.75	> 50	2.0	2.5	0.75	0.1	0.1	0.05	0.2	0.2
	FeV40C0.5	35-48	2.0	0.5	0.5	0.05	0.08	0.03	0.2	2.0
	FeV40C0.75	35-48	2.0	0.5	0.75	0.05	0.08	0.03	0.4	4.0
	FeV40C1	35-48	2.0	0.5	1.0	0.05	0.1	0.03	0.4	6.0

Synthesis

Eighty-five percent of all vanadium extracted from the Earth is used to create alloys such as ferrovanadium. There are two common ways in which ferrovanadium in produced, silicon reduction and aluminum reduction.

Reduction by Silicon

Vanadium pentoxide (V_2O_5), ferrosilicon (FeSi75), lime (CaO) and slag (recycled vanadium containing waste) and are combined in an electric arc furnace heated to 1850 °C. Silicon in the ferrosilicon reduces the vanadium in V_2O_5 to vanadium metal. The vanadium then interacts with the iron to form ferrovanadium. Excess lime and V_2O_5 are added to use up the silicon and refine the metal. This process produces vanadium concentrations between thirty-five and sixty percent.

$$2\ V_2O_5 + 5\ (Fe_{y/5}Si)_{alloy} + 10\ CaO \rightarrow 4\ (Fe_{y/4}V)_{alloy} + 5\ Ca_2SiO_4$$

Reduction by Aluminum

Iron, V_2O_5, aluminum, and lime are combined in an electric arc furnace. Like the silicon, aluminum reduces the vanadium in V_2O_5 to vanadium metal. The vanadium metal dissolves into the iron and forms the ferrovanadium alloy. The resulting ferrovanadium has a vanadium concentration between seventy and eighty-five percent.

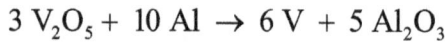

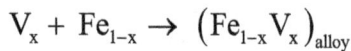

$$3\ V_2O_5 + 10\ Al \rightarrow 6\ V + 5\ Al_2O_3$$
$$V_x + Fe_{1-x} \rightarrow (Fe_{1-x}V_x)_{alloy}$$

Toxicology

Ferrovanadium dust is a mild irritant that affects the eyes when touched by contaminated skin and the respiratory tract when inhaled. The dust caused chronic bronchitis and pneumonitis in animals exposed to high concentration ($1000-2000$ mg/m^3) at intervals for two months. However, no such long-term effects have been observed in humans.

Occupational Exposure

The American Conference of Governmental Industrial Hygienists (ACGIH) states that an employee who is working eight hours a day, five days a week, can be exposed to ferrovanadium dust in their place of work at concentrations of up to 1.0 mg/m^3 without adverse effects. Short-term exposures should be kept below 3.0 mg/m^3. It is suggested that those working with high concentrations of ferrovanadium dust wear a respirator to prevent inhalation and irritation of the respiratory tract.

Steel

The most common use of ferrovanadium is in the production of steel. In 2017, 94% of domestic consumption of vanadium was to produce iron and steel alloys. Ferrovanadium and other vanadium alloys are used in carbon steel, alloy steel high strength steel, and HSLA (High Strength Low Alloy) steel. These steels are then used to make automotive parts, pipes, tools, and more.

The addition of ferrovanadium toughens the steel making it more resistant to temperature and torsion. This increase in strength is a result of the formation of vanadium carbides which have a rigid crystal structure as well as a finer grain size which decreases the ductility of the steel. In addition to adding to the composition of the steel, ferrovanadium can also be used as a coating on the steel. When coated with nitrated ferrovanadium, the abrasion resistance of steel increases 30-50%.

Ferroniobium

Ferroniobium is an important iron niobium alloy, with a niobium content of 60-70%.

Ferroniobium is the main source for niobium alloying of HSLA (high-strength low-alloy) steel. For alloying with steel the ferroniobium is added to molten steel before casting. Since Ferro Niobium can effectively double the strength and toughness, as well as reduce the weight of the alloy, it is a highly desirable compound. Ferro Niobium is an additive to the production process of amorphous metals, and will impart several desirable properties upon the resulting compound. One of the primary benefits of adding Ferro Niobium to an alloy is in its anti-corrosive properties (better than carbon steel). Additionally, the adding of Ferro Niobium to an alloy can make it more weldable and much stronger.

Properties

Physical State	Metallic solid
Colour	Silvered gray
Odour	Odourless
Melting Point	1900°C (FeMo 70%)
Specific Gravity	8.0g/cm³

The product is stable under normal conditions. It may react with acids and alkalis, releasing hydrogen. The dust may ignite when it is suspended in air but a low pressure eliminates any explosion hazard.

Its preparation is not classified as dangerous according to the relevant European regulations, and it is also not classed as a hazardous good for transportation.

Uses

Ferro Niobium is used mainly to increase the strength of the alloys, including:

- The manufacturing of High-Strength special Steels Low-Alloy (HSLA), mainly used in automobiles, pipe fabrication, structural steels and stainless steels resistant to high temperatures.

- The manufacturing of stainless steels.

- The manufacturing of superalloys.

- The increase in weldability, etc.

Ferrocerium

Ferrocerium is a man-made metallic material that gives off a large number of hot sparks at temperatures at 3,000 °F when scraped against a rough surface, such as ridged steel. Because of this property it is used in many applications, such as clockwork toys, strikers for welding torches, so-called "flint-and-steel" or "flint spark lighter" fire-starters in emergency survival kits, and cigarette lighters, as the initial ignition source for the primary fuel. It is also commonly called ferro rod and most commonly of all, mistakenly, flint. As tinder-igniting campfire starter rods it is sold under such trade names as Blastmatch, Fire Steel, and Metal-Match for survivalists and bushcraft hobbyists. Some manufacturers and resellers mistakenly call them "magnesium" rods. It is also known in Europe as Auermetall after its inventor Baron Carl Auer von Welsbach.

Ferrocerium is a synthetic pyrophoric alloy that produces hot sparks that can reach temperatures of 3,000 °C (5,430 °F) when rapidly oxidized by the process of striking. This property allows it to have many commercial applications, such as the ignition source for lighters (where it is often known by the misleading name "flint"), strikers for gas welding and cutting torches, deoxidization in metallurgy, and ferrocerium rods (also called ferro rods, flint-spark-lighters and wrongly "flint-and-steel" as this is the name of a different type of lighter using a section of high carbon steel and a natural flint). Due to ferrocerium's ability to ignite in adverse conditions, rods of ferrocerium are commonly used as an emergency combustion device in survival kits.

Ferrocerium was invented in 1903 by the Austrian chemist Carl Auer von Welsbach. It takes its name from its two primary components: iron (from Latin: *ferrum*), and the rare-earth element cerium. The pyrophoric effect is dependent on the brittleness of the alloy and its low autoignition temperature.

Spark trails from a cigarette lighter

Use

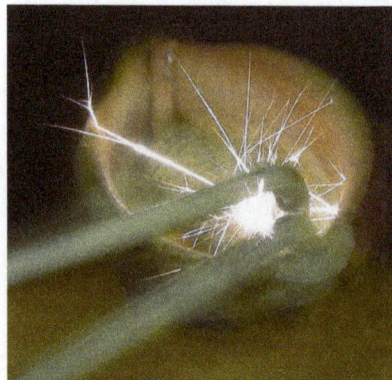

A flint spark lighter in action

While ferrocerium-and-steels function in a similar way to natural flint-and-steel in fire starting, ferrocerium takes on the role that steel played in traditional methods: when small shavings of it are removed quickly enough the heat generated by friction is enough to ignite those shavings, converting the metal to the oxide, i.e., the sparks are tiny pieces of burning metal. The sparking is due to cerium's low ignition temperature of between 150 and 180 °C (302 and 356 °F). About 700 tons were produced in 2000.

Composition

It is also known in Europe as Auermetall after its inventor Baron Carl Auer von Welsbach. Three different Auermetalls were developed: the first was iron and cerium, the second also included lanthanum to produce brighter sparks, and the third added other heavy metals. In the Baron von Welsbach's first alloy, 30% iron (ferrum) was added to purified cerium, hence the name "ferro-cerium".

A modern ferrocerium firesteel product is composed of an alloy of rare-earth metals called mischmetal (containing approximately 20.8% iron, 41.8% cerium, about 4.4% each of praseodymium, neodymium, and magnesium, plus 24.2% lanthanum. A variety of other components are added to modify the spark and processing characteristics. Most contemporary flints are hardened with iron oxide and magnesium oxide.

Element	Iron	Cerium	Neodymium	Praseodymium	Magnesium	Lanthanum
Fraction	20.8%	41.8%	4.4%	4.4%	4.4%	24.2%

Ferrosilicon

Ferrosilicon is a ferroalloy, an alloy of iron and silicon with an average silicon content between 15 and 90 weight percent. It contains a high proportion of iron silicides. Ferrosilicon is used for manufacture of silicon, corrosion-resistant and high-temperature resistant ferrous silicon alloys, and silicon steel for electromotors and transformer cores. In the manufacture of cast iron, ferrosilicon is used for inoculation of the iron to accelerate graphitization. In arc welding, ferrosilicon can be found in some electrode coatings. Ferrosilicon is used as a source of silicon to reduce metals from their oxides and to deoxidize steel and other ferrous alloys. This prevents the loss of carbon from the molten steel (so called blocking the heat). Ferro Silicon is used in the production of cast iron, as Ferro Silicon can accelerate graphitization. Ferro Silicon replaces the need for ferro manganese, spiegeleisen and calcium silicides in the manufacturing process.

Ferro Silicon has a melting point of 1200°C to 1250°C with a boiling point of 2355°C and Ferro Silicon contains about 2% of calcium and aluminium. Ferro Silicon, as an additive to the production process of ferrous metals, will impart several desirable properties upon the resultant alloy. Some of the primary benefits of adding Ferro Silicon to an alloy is to improve the corrosion resistant properties of the new compound, as well as to add to the high temperature heat-resistance properties of the new alloy, for example, in the production of silicon steel for use in transformer cores.

A large portion of the global Ferro Silicon supply is manufactured in China, USA and India. The most basic definition of the Ferro Silicon production process would be that the silica (or sand) is mixed with coke, and then a reduction process takes place in the presence of millscale, scrap or another source of iron. A blast furnace is employed for Ferro Silicon production, but for larger contents of silica, an electric arc furnace is used.

There are many practical applications of Ferro Silicon to include carbon steel and stainless steel production, and when using the Pidgeon process to produce magnesium from dolomite. Applications in the production of other alloys include the manufacture of silicon steel for electro motors and cores, as well as coatings used during arc welding. One useful by-product of the production processes is silica fume, which is later added to concrete mixes to improve compressive and bonding strength there.

Uses

Ferrosilicon is used as a source of silicon to reduce metals from their oxides and to deoxidize steel and other ferrous alloys. This prevents the loss of carbon from the molten steel (so called *blocking the heat*); ferromanganese, spiegeleisen, Calcium silicides, and many other materials are used for the same purpose. It can be used to make other ferroalloys. Ferrosilicon is also used for manufacture of silicon, corrosion-resistant and high-temperature-resistant ferrous silicon alloys, and silicon steel for electromotors and transformer cores. In the manufacture of cast iron, ferrosilicon is used for inoculation of the iron to accelerate graphitization. In arc welding, ferrosilicon can be found in some electrode coatings.

Ferrosilicon is a basis for manufacture of prealloys like magnesium ferrosilicon (MgFeSi), used for production of ductile iron. MgFeSi contains 3–42% magnesium and small amounts of rare-earth metals. Ferrosilicon is also important as an additive to cast irons for controlling the initial content of silicon.

Magnesium ferrosilicon is instrumental in the formation of nodules, which give ductile iron its flexible property. Unlike gray cast iron, which forms graphite flakes, ductile iron contains graphite nodules, or pores, which make cracking more difficult.

Ferrosilicon is also used in the Pidgeon process to make magnesium from dolomite. Treatment of high-silicon ferrosilicon with hydrogen chloride is the basis of the industrial synthesis of trichlorosilane.

Hydrogen Production

Ferrosilicon is used by the military to quickly produce hydrogen for balloons by the ferrosilicon method. The chemical reaction uses sodium hydroxide, ferrosilicon, and water. The generator is small enough to fit in a truck and requires only a small amount of electric power, the materials are stable and not combustible, and they do not generate hydrogen until mixed. The method has been in use since World War I. Prior to this, the process and purity of hydrogen generation relying on steam passing over hot iron was difficult to control. While in the "silicol" process, a heavy steel pressure vessel is filled with sodium hydroxide and ferrosilicon, and upon closing, a controlled amount of water is added; the dissolving of the hydroxide heats the mixture to about 200 °F (93 °C) and starts the reaction; sodium silicate, hydrogen and steam are produced.

Ferroboron

Ferro Boron is an alloy, which is formed by combining iron and boron. Ferro Boron will typically be added to steel to form a very strong, highly durable and specialist steel for use across a variety of applications. Although the uses of Ferro Boron as a steel additive are limited due to the relative high cost of production, it is still considered to be one of the most useful ferro alloys.

Ferro Boron as an additive to the production process of amorphous metals will impart several desirable properties to the new alloy. One of the primary benefits of adding Ferro Boron to an alloy is in the fact that it can significantly increase the magnetic susceptibility of the final alloy, making it ideal for the production of Nd-Fe-B magnets. Additionally, the adding of Ferro Boron to an alloy can drastically increase the deep quench ability of the alloy which is finally produced. The final property that will be impacted by Ferro Boron when it is added to an alloy is that it will add a significant wash resistance.

A large portion of the global Ferro Boron supply is manufactured in China, India and Turkey. The most basic definition of the Ferro Boron production process would be that a small amount of raw Boron is added to iron to create a ferrous alloy, although the actual methods used are somewhat more complicated including fusing boric acid prior to use. Typically the resulting alloy will be produced as either small nuggets, or as a finer powder, which will be packaged into 50kg, 100kg or even 1000kg units.

Ferro Boron is quite widely used in a variety of applications, although the predominant ones are hot spray painting and in the production of amorphous metals. There have been wide ranging experiments using Ferro Boron to produce prototype steel compounds, many of which were entirely successful, although not suitable for general manufacturing as other forms of steel are cheaper to produce. One of the more advanced uses of Ferro Boron is in the production of magnetic glass, where it will be added to the silicon compound, along with a small quantity of iron, during the initial production process. The most common use of Ferro Boron, aside from the steel industry is in the production of magnets, where it will add significant magnetic susceptibility to the final product.

Chemical Specifications of Ferro Boron

Ferro Boron is an alloy, which is formed by combining iron and boron. Ferro Boron will typically be added to steel to form a very strong, highly durable and specialist steel for use across a variety of applications. Although the uses of Ferro Boron as a steel additive are limited due to the relative high cost of production, it is still considered to be one of the most useful ferro alloys.

Ferro Boron as an additive to the production process of amorphous metals will impart several desirable properties to the new alloy. One of the primary benefits of adding Ferro Boron to an alloy is in the fact that it can significantly increase the magnetic susceptibility of the final alloy, making it ideal for the production of Nd-Fe-B magnets. Additionally, the adding of Ferro Boron to an alloy can drastically increase the deep quench ability of the alloy which is finally produced. The final property that will be impacted by Ferro Boron when it is added to an alloy is that it will add a significant wash resistance.

A large portion of the global Ferro Boron supply is manufactured in China, India and Turkey. The most basic definition of the Ferro Boron production process would be that a small amount of raw Boron is added to iron to create a ferrous alloy, although the actual methods used are somewhat more complicated including fusing boric acid prior to use. Typically the resulting alloy will be produced as either small nuggets, or as a finer powder, which will be packaged into 50kg, 100kg or even 1000kg units.

Ferro Boron is quite widely used in a variety of applications, although the predominant ones are hot spray painting and in the production of amorphous metals. There have been wide ranging experiments using Ferro Boron to produce prototype steel compounds, many of which were entirely successful, although not suitable for general manufacturing as other forms of steel are cheaper to produce. One of the more advanced uses of Ferro Boron is in the production of magnetic glass, where it will be added to the silicon compound, along with a small quantity of iron, during the initial production process. The most common use of Ferro Boron, aside from the steel industry is in the production of magnets, where it will add significant magnetic susceptibility to the final product.

Chemical Specifications of Ferro Boron

Element	MAC-FeB-14
B	14 % min
C	1.5 % max
Si	2 % max
Al	1 % max
S	0.02 % max
P	0.10 % max

Properties

Physical State	Metallic solid
Colour	Silver Metallic Grey
Odour	Odourless
Melting Point	>1500°C
Boiling Point	-
Specific Gravity	3.3g/cm³

The product is stable under normal storage and handling conditions. Exposure to moisture should be avoided due to the risk involved in adding a wet or moist product to a molten bath. The powder product is not compatible with chlorine, oxidizing agents, strong acids and polystyrene. Under certain conditions it can form boron oxide and boron fluoride.

This product is neither classified as hazardous according to the relevant European regulations nor as a hazardous good for its transportation.

Uses

The uses of Ferro Boron include:

- The improvement of hardenability of low alloyed steels.

- The boration surface treatment of steels.

- The reduction of nitrogen.

- The manufacturing of NdFeB permanent magnets.

- The manufacturing of metallic glass.

Ferromolybdenum

Ferromolybdenum is an alloy formed by combining iron and molybdenum. It is an extremely versatile alloy used primarily in highstrength low alloys and stainless steels. It has numerous beneficial properties and can be used even in cast irons, some high-speed tool steels, and superalloy applications. Adding ferromolybdenum to a material helps to improve weldability, corrosion and wear resistance as well to increase ferrite strength.

Applications

The largest application area of ferromolybdenum is in the manufacture of ferrous alloys. Based on the range of molybdenum content, ferromolybdenum can be applied in the manufacture of machine tools and equipment, military hardware, refinery tubing, load-bearing parts and rotary drills.

Ferromolybdenum is also used in cars, trucks, locomotives and ships. Ferromolybdenum is added to stainless and heat-resisting steels that are used in synthetic fuel and chemical plants, heat exchangers, power generators, oil-refining equipment, pumps, turbine tubing, ship propellers, plastics and inside acid storage containers.

Uses of Ferro Molybdenum

The largest practical applications of Ferro Molybdenum are its use in ferrous alloys, and depending on the molybdenum content range, it is suited for machine tools and equipment, military hardware, refinery tubing, load-bearing parts and rotary drills. Ferro Molybdenum is also used in cars, trucks, locomotives and ships. In addition, Ferro Molybdenum is used in stainless and heat-resisting steels that are employed by synthetic fuel and chemical plants, heat exchangers, power generators, oil-refining equipment, pumps, turbine tubing, ship propellers, plastics and inside acid storage containers. Tool steels, with a high percentage range of Ferro Molybdenum, are used in highspeed machining parts, cold work tools, drill bits, screwdrivers, dies, chisels, heavy castings, ball and rolling mills, rolls, cylinder blocks, piston rings and large drill bits.

Ferromolybdenum can be used in any melting process to add molybdenum to all types of iron and steel, and is supplied in a range of sizes for furnace or ladle addition. The recovery should be substantially 100 % if used correctly. For optimum recoveries with ladle additions, ferromolybdenum should be added after the molten metal has covered the bottom of the ladle and before it is three quarters full.

Maraging Steels

The maraging steels offer a remarkable combination of strength and ductility, with a yield strength as high as 2400 MPa with total elongation of 6%. By virtue of their high cost, maraging steels are used mainly in special applications such as rocket casing and other aerospace applications.

In 1958, Bieber showed that low carbon iron-nickel martensites containing titanium and aluminium could be precipitation hardened to HRC-65. Maraging steels are highly alloyed low carbon iron-nickel martensites which possess an excellent combination of strength and toughness better than most carbon-hardened steels. It is an attempt to replace carbon steels at least in critical applications.

High strength in maraging steels is developed due to the formation of very low carbon, tough and ductile iron-nickel martensite but comparatively soft, about 30 HRc, which is further strengthened by the precipitation of intermetallic compounds during age hardening. These steels have high fracture toughness K_{IC} = 120 MNm$^{-3/2}$. As maraging steels are practically non-carbon (C max. 0.03%), no carbides precipitate here.

Composition of Maraging Steels

There is practically no carbon in them (less than 0.01-0.03%). A nickel content of around 18% helps to get complete lath type martensite formed by air cooling to room temperature. A larger volume of finely dispersed intermetallic compounds which induce high yield strength in maraging steels require nickel content of around 18%. Presence of more than 23% Nickel results in undesirable twinned martensite.

To ensure high toughness, the contents of residual elements P, S, C and N are kept as low as possible otherwise, segregation or precipitation mainly of Ti (C, N) and Ti_2S at prior austenite grain boundary causes embrittlement. Embrittled material can be restored to ductile state (if large % of elements are present) by a solution treatment at about 980°C followed by rapid cooling.

Molybdenum (≈ 6%) and titanium (≈ max 2%) are necessary additions as these form hardening precipitates Ni_3Mo, Ni_3Ti and even Lave phase, Fe_2Mo. Titanium is a stronger hardener, molybdenum and aluminium are moderate hardeners, and cobalt (does not form the precipitates) is a weak hardener. Higher than 1.2% titanium reduces ductility before and after ageing. Molybdenum forms rod shaped Ni_3Mo (25A° wide and 500 A° long at peak hardness) coherent precipitates, which distort the BCC matrix to increase strength and hardness.

Titanium forms stable Ni_3Ti and distorts the BCC lattice. Molybdenum reduces grain boundary precipitation to avoid severe drop in ductility. The presence of more than 1% Mo promotes undesirable austenite reversion. The yield strength of 18% Ni-8% Co-5% Mo increases from 1375 MPa to 2410 MPa as titanium contents increases from 0.20% to 1.4%.

Cobalt has many roles to play. To obtain complete lath type martensite on cooling to room temperature, M_s temperature should be 200° to 300°C. Additions of all the important elements in iron lower the M_s temperature.

One of the roles of 6-8% cobalt in maraging steel in to raise M_s temperature, so that greater amounts of titanium, molybdenum can be added as hardeners and still allow complete transformation to lath martensite before the steels cools to room temperature.

The absence of cobalt in maraging steels requires reduction of these elements and even nickel is then not present higher than 18%. Cobalt does not form precipitate but increases hardness and strength of maraging steels by forming short-range ordering.

Cobalt has greater affinity for iron, and thus does not allow easily metastable. Ni-rich precipitates to change to equilibrium iron-rich precipitate Fe_2Mo which reduces the hardness. Cobalt also retards reversion to undesirable austenite.

Cobalt reduces solid solubility of molybdenum in BCC matrix and promotes fine dispersion of Ni_3Mo precipitate particles and increases the volume fraction of this precipitate. Aluminium is present up to 0.1% as it also raises M_s temperature.

Table: shows some maraging steels with their properties.

	Element, Wt %					Yield strength MPa	Tensile strength MPa	% Elongation (50 mm)	Reduction in Area %	Heat Treatment
	Ni	Mo	Co	Ti	Al					
200	18	3.3	8.5	0.2	0.1	1400	1500	10	60	A*
250	18	5.0	7.75	0.4	0.1	1700	1800	8	55	A
300	18	5.0	9.00	0.65	0.1	2000	2050	7	40	A
350	18	4.2	12.50	1.6	0.1	2400	2450	6	25	B
Cobalt-free										

200	18.5	3.0	-	0.7	0.1					
250	18.5	3.0	-	1.4	0.1	1825	1895	11.5	58.5	C
300	18.5	4.0	-	1.85	0.1					

Characteristics of Maraging Steels

1. As there is practically no carbon in them, martensite is soft, tough, ductite and readily machinable. Ageing is done after machining operations. No carbides are precipitated.

2. As the CCT diagrams of 18 Ni maraging steels don't show presence of bainite and pearlite transformations, these steels transform to martensite on air cooling from austenitic range. The problems of distortion thus, are almost non-existent.

3. Of the two types of martensites, alloy composition of maraging steels is so chosen to get complete transformation to lath martensite as the steel cools to room temperature. Lath martensite has high density of dislocations, which are uniformly distributed. These promote improved response to age-hardening, first by providing a large number of preferred nucleation sites for the inter-metallic compound precipitates, to form during ageing; secondly, these provide preferred diffusion paths (during ageing) for substitutional solute atoms. Thus, ageing is done for 3-5 hours at 480°c. Ageing results in fine dispersion of precipitates with no preference for grain boundaries.

4. The Fe-Ni maraging steels exhibit athermal hysteresis in phase transformation, i.e., on heating for ageing, the reversion of martensite to austenite is prevented, or minimised.

5. Retained austenite causes variations in tensile strength, ductility, and toughness and is thus undesirable. Twinned martensite is also not desirable. Thus, M_s temperature of the steels is kept high around 200° to 300°C.

6. Normally, maraging steels ingots are heated to 1250°C with hot rolling finished at about 920°C. The steels are subsequently reheated to 815°C to austenitise them. Lower austenitising temperatures result in lower strength and ductility due to incomplete solution of hardening elements. Higher temperatures result in grain growth to reduce the strength of the steels. Boron 0.001 to 0.003% minimises grain growth and resulting reduction in strength. Normal austenitising time, or called as solution-annealing time is 1 hr.

7. Thermal cycling of these steels between M_f and a temperature above solution-annealing temperature does refine the grain size by recrystallisation, but grain size finer than ASTM 6 or 7 cannot be achieved by this method.

8. Age hardening is normally done at 455 to 510°C for 3-12 hours, and then air cooled to room temperature. Normally 18 Ni steels are aged at 480°C for 3 hrs.

    Figure illustrates effect of ageing 18 Ni 250 alloy at different ageing temperatures. Overaging appears to improve significantly the plain-strain fracture toughness (K_{IC}).

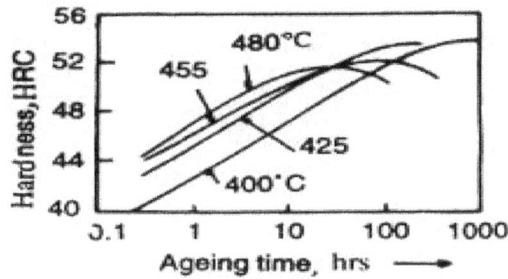

Effect of aging at different temperatures of 18 Ni 250 steel after solution annealing at 815°C for 1 hr.

9. During ageing, first short range ordering occurs to form (i) iron and cobalt rich regions, (ii) nickel-rich regions. Ni_3 Mo (rod-shape), Ni_3 Ti compounds are formed in nickel-rich regions. These precipitates form on dislocations and the lath boundaries, i.e., are uniformly and finely formed to increase the hardness. Short range order of cobalt also increases hardness. The size of these precipitates increases to ultimately attain the peak hardness. The coherency strains become large to help in dissolving Ni_3Mo and to precipitate Fe_2Mo particles. This is the overageing. Simultaneously, reversion of austenite also occurs. These factors result in decrease in hardness.

10. Nitriding can be done of maraging steels to induce hardness 65-70 HRC up to a depth of 0.15 mm after nitriding for 24-48 hours at 455°C. This improves wear resistance and fatigue resistance.

11. During age hardening, no incubation period is needed for the precipitation, as here is no free energy barrier to precipitation. This is because there is high degree of super saturation in the alloy, and precipitation occurs due to heterogeneous nucleation on the dislocations. The diffusion paths are available of dislocations for solutes to diffuse easily and quickly to cause precipitation soon.

12. Maraging steels are easily weldable. Conventional welding techniques produce sound welds. However, after welding, solution annealing and ageing may be done.

References

- Iron-alloy: themetalcasting.com, Retrieved 10 March, 2019

- Iron-alloys, materials-science: sciencedirect.com, Retrieved 2 July, 2019

- Iron-alloy: themetalcasting.com, Retrieved 13 May, 2019

- Cast-iron, technology: britannica.com, Retrieved 17 January, 2019

- Krause, Keith (August 1995). Arms and the State: Patterns of Military Production and Trade. Cambridge University Press. p. 40. ISBN 978-0-521-55866-2

- Glossary of Metalworking Terms. Industrial Press. 2003. p. 297. ISBN 9780831131289. Archived from the original on 2017-02-24

- Cementite, Lattices: phase-trans.msm.cam.ac.uk, Retrieved 7 February, 2019

- Carroll, P. K.; P. McCormack (1 October 1972). "The Spectrum of FeH: Laboratory and Solar Identification". Astrophysical Journal Letters. 177: L33–L36. Bibcode:1972ApJ...177L..33C. doi:10.1086/181047.

- Ferroalloy, technology: britannica.com, Retrieved 13 April, 2019

- Ferroalloys, materials-science: sciencedirect.com, Retrieved 3 June, 2019

- Ferrosilicon, ferro-alloys: metalandalloyscorporation.com, Retrieved 18 May, 2019

- Ramesh Singh (3 October 2011). Applied Welding Engineering: Processes, Codes, and Standards. Elsevier. pp. 38–. ISBN 978-0-12-391916-8. Retrieved 25 December 2011

- Ferroboron, ferro-alloys: metalandalloyscorporation.com, Retrieved 8 May, 2019

- Maraging-steels-composition-and-characteristics-metallurgy, steel, metallurgy: engineeringenotes.com, Retrieved 10 August, 2019

Understanding Steel

Steel refers to an alloy of iron with carbon and other elements. There are numerous types of steel such as carbon steel, alloy steel, tool steel, stainless steel and structural steel. Carbon steel has three major sub-types, namely, low carbon, medium carbon and high carbon steel. This chapter discusses in detail these types of steel as well as the processes involved in the treatment of steel.

Steel is an alloy of iron and carbon in which the carbon content ranges up to 2 percent (with a higher carbon content, the material is defined as cast iron). By far the most widely used material for building the world's infrastructure and industries, it is used to fabricate everything from sewing needles to oil tankers. In addition, the tools required to build and manufacture such Products are also made of steel. As an indication of the relative importance of this material, in 2013 the world's raw steel production was about 1.6 billion tons, while production of the next most important engineering metal, aluminum, was about 47 million tons. The main reasons for the popularity of steel are the relatively low cost of making, forming, and processing it, the abundance of its two raw materials (iron ore and scrap), and its unparalleled range of mechanical properties.

Fundamentally, all steels are mixtures, or more properly, alloys of iron and carbon. However, even the so-called plain-carbon steels have small, but specified, amounts of manganese and silicon plus small and generally unavoidable amounts of phosphorus and sulfur. The carbon content of plain-carbon steels may be as high as 2.0%, but such an alloy is rarely found. Carbon content of commercial steels usually ranges from 0.05 to about 1.0%.

The alloying mechanism for iron and carbon is different from the more common and numerous other alloy systems in that the alloying of iron and carbon occurs as a two-step process. In the initial step, iron combines with 6.67% C, forming iron carbide, which is called cementite. Thus, at room temperature, conventional steels consist of a mixture of cementite and ferrite (essentially iron). Each of these is known as a phase (defined as a physically homogeneous and distinct portion of a material system). When a steel is heated above 725 C (1340 F), cementite dissolves in the matrix, and a new phase is formed, which is called austenite. Note that phases of steel should not be confused with structures. There are only three phases involved in any steel—ferrite, carbide (cementite), and austenite, whereas there are several structures or mixtures of structures.

Steel, however, is by far the most widely used alloy and for a very good reason. Among layman, the reason for steel's dominance is usually considered to be the abundance of iron ore (iron is the principal ingredient in all steels) and/or the ease by which it can be refined from ore. Neither of these is necessarily correct; iron is by no means the most abundant element, and it is not the easiest metal to produce from ore. Copper, for example, exists as nearly pure metal in certain parts of the world.

Steel is such an important material because of its tremendous flexibility in metal working and heat treating to produce a wide variety of mechanical, physical, and chemical properties.

Metallurgical Phenomena

The broad possibilities provided by the use of steel are attributed mainly to two all-important metallurgical phenomena: iron is an allotropic element; that is, it can exist in more than one crystalline form; and the carbon atom is only 1/30 the size of the iron atom. These phenomena are thus the underlying principles that permit the achievements that are possible through heat treatment.

In entering the following discussion of constitution, however, it must be emphasized that a maximum of technical description is unavoidable. This portion of the subject is inherently technical. To avoid that would result in the discussion becoming uninformative and generally useless. The purpose of this chapter is, therefore, to reduce the prominent technical features toward their broadest generalizations and to present those generalizations and underlying principles in a manner that should instruct the reader interested in the metallurgical principles of steel. This is done at the risk of some oversimplification.

Constitution of Iron

It should first be made clear to the reader that any mention of molten metal is purely academic; this book deals exclusively with the heat treating range that is well below the melting temperature. The objective of this section is to begin with a generalized discussion of the constitution of commercially pure iron, subsequently leading to discussion of the ironcarbon alloy system that is the basis for all steels and their heat treatment.

All pure metals, as well as alloys, have individual constitutional or phase diagrams. As a rule, percentages of two principal elements are shown on the horizontal axis of a figure, while temperature variation is shown on the vertical axis. However, the constitutional diagram of a pure metal is a simple vertical line. The constitutional diagram for commercially pure iron is presented in figure. This specific diagram is a straightline as far as any changes are concerned, although time is indicated on the horizontal. As pure iron, in this case, cools, it changes from one phase to another at constant temperature. No attempt is made, however, to quantify time, but merely to indicate as a matter of interest that as temperature increases, reaction time decreases, which is true in almost any solidsolution reaction.

Pure iron solidifies from the liquid at 1538 °C (2800 °F). A crystalline structure, known as ferrite, or delta iron, is formed (point a, figure). This structure, in terms of atom arrangement, is known as a body-centered cubic lattice (bcc), shown in figure. This lattice has nine atoms—one at each corner and one in the center.

As cooling proceeds further and point b is reached (1395 °C, or 2540 °F), the atoms rearrange into a 14-atom lattice a shown in figure.

The lattice now has an atom at each corner and one at the center of each face. This is known as a face-centered cubic lattice (fcc), and this structure is called gamma iron.

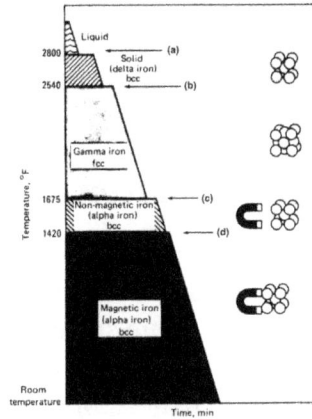

Changes in pure iron as it cools from the molten state to room temperature.

As cooling further proceeds to 910 ° C (1675 ° F), the structure reverts to the nine-atom lattice or alpha iron. The change at point d on figure. (770 ° C, or 1420 ° F) merely denotes a change from non-magnetic to magnetic iron and does not represent a phase change. The entire field below 910 ° C (1675 ° F) is composed of alpha ferrite, which continues on down to room temperature and below. The ferrite forming above the temperature range of austenite is often referred to as delta ferrite; that forming below A3 as alpha ferrite, though both are structurally similar. In this Greek-letter sequence, austenite is gamma iron, and the interchangeability of these terms should not confuse the fact that only two structurally distinct forms of iron exist.

Figures thus illustrate the allotropy of iron. In the following sections of this chapter, the mechanism of allotropy as the all-important phenomenon relating to the heat treatment of iron-carbon alloys is discussed.

Alloying Mechanisms

Metal alloys are usually formed by mixing together two or more metals in their molten state. The two most common methods of alloying are by atom exchange and by the interstitial mechanism. The mechanism by which two metals alloy is greatly influenced by the relative atom size. The exchange mechanism simply involves trading of atoms from one lattice system to another. An example of alloying by exchange is the coppernickel system wherein atoms are exchanged back and forth.

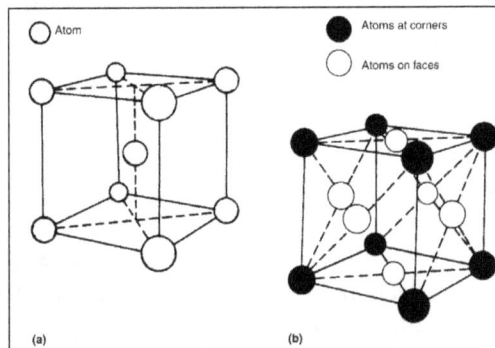

Arrangement of atoms in the two crystalline structures of pure iron.
(a) Body-centered cubic lattice. (b) Face-centered cubic lattice.

Interstitial alloying requires that there be a large variation in atom sizes between the elements involved. Because the small carbon atom is 1/30 the size of the iron atom, interstitial alloying is easily facilitated. Under certain conditions, the tiny carbon atoms enter the lattice (the interstices) of the iron crystal. A description of this basic mechanism follows.

Effect of Carbon on the Constitution of Iron

As an elemental metal, pure iron has only limited engineering usefulness despite its allotropy. Carbon is the main alloying addition that capitalizes on the allotropic phenomenon and lifts iron from mediocrity into the position of a unique structural material, broadly known as steel. Even in the highly alloyed stainless steels, it is the quite minor constituent carbon that virtually controls the engineering properties. Furthermore, due to the manufacturing processes, carbon in effective quantities persists in all irons and steels unless special methods are used to minimize it.

Carbon is almost insoluble in iron, which is in the alpha or ferritic phase (910 ° C, or 1675 ° F). However, it is quite soluble in gamma iron. Carbon actually dissolves; that is, the individual atoms of carbon lose themselves in the interstices among the iron atoms. Certain interstices within the fcc structure (austenite) are considerably more accommodating to carbon than are those of ferrite, the other allotrope. This preference exists not only on the mechanical basis of size of opening, however, for it is also a fundamental matter involving electron bonding and the balance of those attractive and repulsive forces that underlie the allotrope phenomenon.

The effects of carbon on certain characteristics of pure iron are shown in figure. Figure is a simplified version of figure. that is, a straight line constitutional diagram of commercially pure iron. In figure, the diagram is expanded horizontally to depict the initial effects of carbon on the principal thermal points of pure iron. Thus, each vertical dashed line, like the solid line in figure, is a constitutional diagram, but now for iron containing that particular percentage of carbon. Note that carbon lowers the freezing point of iron and that it broadens the temperature range of austenite by raising the temperature A_4 at which (delta) ferrite changes to austenite and by lowering the temperature A_3 at which the austenite reverts to (alpha) ferrite. Hence, carbon is said to be an austenitizing element. The spread of arrows at A_3 covers a two-phase region, which signifies that austenite is retained fully down to the temperatures of the heavy arrow, and only in part down through the zone of the lesser arrows.

In a practical approach, however, it should be emphasized that figure, as well as figure represents changes that occur during very slow cooling, as would be possible during laboratory-controlled experiments, rather than under conditions in commercial practice. Furthermore, in slow heating of iron, these transformations take place in a reverse manner. Transformations occurring at such slow rates of cooling and heating are known as equilibrium transformations, due to the fact that temperatures indicated in figure.

Therefore, the process by which iron changes from one atomic arrangement to another when heated through 910 ° C (1675 ° F) is called a transformation. Transformations of this type occur not only in pure iron but also in many of its alloys; each alloy composition transforms at its own characteristic temperature. It is this transformation that makes possible the variety of properties that can be achieved to a high degree of reproducibility through use of carefully selected heat treatments.

Iron-Cementite Phase Diagram

When carbon atoms are present, two changes occur. First, transformation temperatures are lowered, and second, transformation takes place over a range of temperatures rather than at a single temperature. These data are shown in the well-known iron-cementite phase diagram. However, a word of explanation is offered to clarify the distinction between phases and phase diagrams.

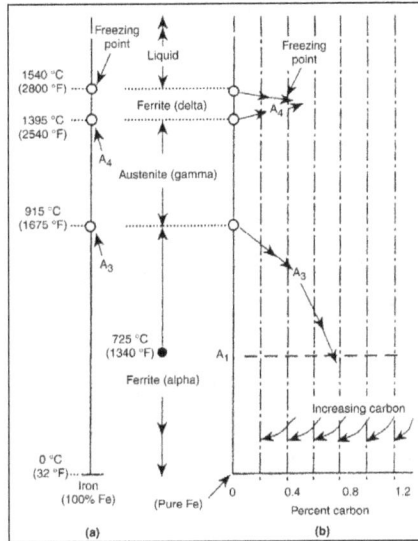

Effects of carbon on the characteristics of commercially pure iron. (a) Constitutional diagram for pure iron. (b) Initial effects of carbon on the principal thermal points of pure iron.

A phase is a portion of an alloy, physically, chemically, or crystallographically homogeneous throughout, which is separated from the rest of the alloy by distinct bounding surfaces. Phases that occur in iron-carbon alloys are molten alloy, austenite (gamma phase), ferrite (alpha phase), cementite, and graphite. These phases are also called constituents. Not all constituents (such as pearlite or bainite) are phases—these are microstructures.

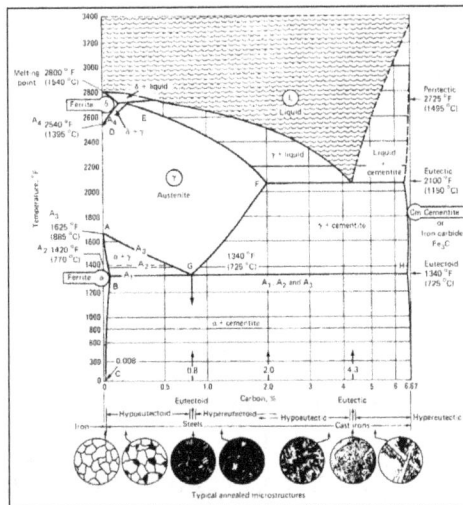

Iron-cementite phase diagram.

A phase diagram is a graphical representation of the equilibrium temperature and composition limits of phase fields and phase reactions in an alloy system. In the iron-cementite system, temperature is plotted vertically, and composition is plotted horizontally. The iron-cementite diagram, deals only with the constitution of the iron-iron carbide system, i.e., what phases are present at each temperature and the composition limits of each phase. Any point on the diagram, therefore, represents a definite composition and temperature, each value being found by projecting to the proper reference axis.

Although this diagram extends from a temperature of 1870 °C (3400 °F) down to room temperature, note that part of the diagram lies below 1040 °C (1900 °F). Steel heat treating practice rarely involves the use of temperatures above 1040 °C (1900 °F). In metal systems, pressure is usually considered as constant.

Frequent reference is made to the iron-cementite diagram in this chapter and throughout this book. Consequently, understanding of this concept and diagram is essential to further discussion.

The iron-cementite diagram is frequently referred to incorrectly as the iron-carbon equilibrium diagram. Iron-"carbon" is incorrect because the phase at the extreme right is cementite, rather than carbon or graphite; the term equilibrium is not entirely appropriate because the cementite phase in the iron-graphite system is not really stable. In other words, given sufficient time (less is required at higher temperatures), iron carbide (cementite) decomposes to iron and graphite, i.e., the steel graphitizes. This is a perfectly natural reaction, and only the iron-graphite diagram is properly referred to as a true equilibrium diagram.

Solubility of Carbon in Iron

In figure above, the area denoted as austenite is actually an area within which iron can retain much dissolved carbon. In fact, most heat treating operations (notably annealing, normalizing, and heating for hardening) begin with heating the alloy into the austenitic range to dissolve the carbide in the iron. At no time during such heating operations are the iron, carbon, or austenite in the molten state. A solid solution of carbon in iron can be visualized as a pyramidal stack of basketballs with golf balls between the spaces in the pile. In this analogy, the basketballs would be the iron atoms, while the golf balls interspersed between would be the smaller carbon atoms.

Austenite is the term applied to the solid solution of carbon in gamma iron, and, like other constituents in the diagram, austenite has a certain definite solubility for carbon, which depends on the temperature. As indicated by the austenite area in figure. The carbon content of austenite can range from 0 to 2%. Under normal conditions, austenite cannot exist at room temperature in plain carbon steels; it can exist only at elevated temperatures bounded by the lines AGFED in figure. Although austenite does not ordinarily exist at room temperature in carbon steels, the rate at which steels are cooled from the austenitic range has a profound influence on the room temperature microstructure and properties of carbon steels. Thus, the phase known as austenite is fcc iron, capable of containing up to 2% dissolved carbon.

The solubility limit for carbon in the bcc structure of iron-carbon alloys is shown by the line ABC in figure. This area of the diagram is labeled alpha (α), and the phase is called ferrite. The maximum solubility of carbon in alpha iron (ferrite) is 0.025% and occurs at 725°C (1340°F). At room temperature, ferrite can dissolve only 0.008% C, as shown in figure. This is the narrow area at the extreme left of Figure. below approximately 910 °C (1675 °F). For all practical purposes, this area has no effect on heat treatment and shall not be discussed further. Further discussion of figure is necessary, although as previously stated, the area of interest for heat treatment extends vertically to only about 1040 °C (1900 °F) and horizontally to a carbon content of 2%. The large area extending vertically from zero to the line BGH (725 °C, or 1340 °F) and horizontally to 2% C is denoted as a two-phase area— α + Cm, or alpha (ferrite) plus cementite (carbide). The line BGH is known as the lower transformation temperature (A1). The line AGH is the upper transformation temperature (A3). The triangular area ABG is also a two-phase area, but the phases are alpha and gamma, or ferrite plus austenite. As carbon content increases, the A3 temperature decreases until the eutectoid is reached— 725 °C (1340 °F) and 0.80% C (point G). This is considered a saturation point; it indicates the amount of carbon that can be dissolved at 725 °C (1340 °F). A1 and A3 intersect and remain as one line to point H as indicated. The area above 725 °C (1340 °F) and to the right of the austenite region is another two-phase field—gamma plus cementite (austenite plus carbide).

Now as an example, when a 0.40% carbon steel is heated to 725 °C (1340 °F), its crystalline structure begins to transform to austenite; transformation is not complete however until a temperature of approximately 815 °C (1500 °F) is reached. In contrast, as shown in figure a steel containing 0.80% C transforms completely to austenite when heated to 725 °C (1340 °F). Now assume that a steel containing 1.0% C is heated to 725 °C (1340 °F) or just above. At this temperature, austenite is formed, but because only 0.80% C can be completely dissolved in the austenite, 0.20% C remains as cementite, unless the temperature is increased. However, if the temperature of a 1.0% carbon steel is increased above about 790 °C (1450 °F), the line GF is intersected, and all of the carbon is thus dissolved. Increasing temperature gradually increases the amount of carbon that can be taken into solid solution. For instance, at 1040 °C (1900 °F), approximately 1.6% C can be dissolved.

Transformation of Austenite

Thus far the discussion has been confined to heating of the steel and the phases that result from various combinations of temperature and carbon content. Now what happens when the alloy is cooled? Referring to figure, assume that a steel containing 0.50% C is heated to 815 °C (1500 °F). All of the carbon will be dissolved (assuming, of course, that holding time is sufficient). Under these conditions, all of the carbon atoms will dissolve in the interstices of the fcc crystal. If the alloy is cooled slowly, transformation to the bcc or alpha phase begins when the temperature drops below approximately 790 °C (1450 °F). As the temperature continues to decrease, the transformation is essentially complete at 725 °C (1340 °F). During this transformation, the carbon atoms escape from the lattice because they are essentially insoluble in the alpha crystal (bcc). Thus, in slow cooling, the alloy for all practical purposes, returns to the same state (in terms of phase) that it was before heating to form austenite. The same mechanism occurs with higher carbon steels, except that the austenite-to-ferrite transformation does not go through a two-phase zone. In addition to the entry and exit of the carbon atoms through the interstices of the iron atoms, other changes occur that affect the practical aspects of heat treating.

First, a magnetic change occurs at 770 °C (1420 °F) as shown in figure. The heat of transformation effects may chemical changes, such as the heat that is evolved when water freezes into ice and the heat that is absorbed when ice melts. When an iron-carbon alloy is converted to austenite by heat, a large absorption of heat occurs at the transformation temperature. Likewise, when the alloy changes from gamma to alpha (austenite to ferrite), heat evolves.

What happens when the alloy is cooled rapidly? When the alloy is cooled suddenly, the carbon atoms cannot make an orderly escape from the iron lattice. This cause "atomic bedlam" and results in distortion of the lattice, which manifests itself in the form of hardness and/or strength. If cooling is fast enough, a new structure known as martensite is formed, although this new structure (an aggregate of iron and cementite) is in the alpha phase.

Heat Treating of Steel

Heat treatment of steels is the heating and cooling of metals to change their physical and mechanical properties, without letting it change its shape. Heat treatment could be said to be a method for strengthening materials but could also be used to alter some mechanical properties such as improving formability, machining, etc. The most common application is metallurgical but heat treatment of metals can also be used in the manufacture of glass, aluminum, steel and many more materials.

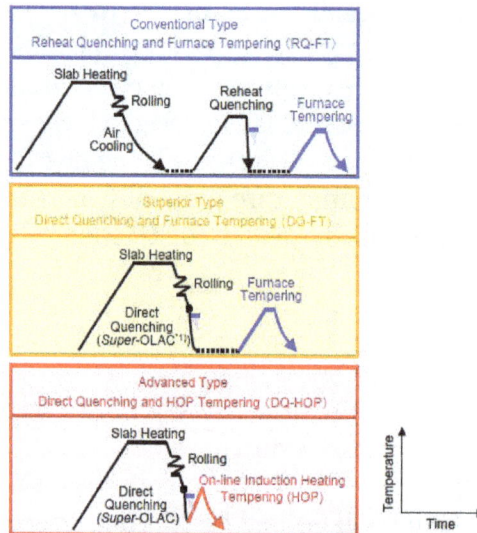

Heat Treatment of Steels.

The process of heat treatment involves the use of heating or cooling, usually to extreme temperatures to achieve the desired result. It is a very important manufacturing processes that can not only help the manufacturing process but can also improve product, its performance, and its characteristics in many ways.

Heat Treatment Processes

Hardening

Hardening involves heating of steel, keeping it at an appropriate temperature until all pearlite is transformed into austenite, and then quenching it rapidly in water or oil. The temperature at which austentizing rapidly takes place depends upon the carbon content in the steel used. The heating time should be increased ensuring that the core will also be fully transformed into austenite. The microstructure of a hardened steel part is ferrite, martensite, or cementite.

Tempering

Tempering involves heating steel that has been quenched and hardened for an adequate period of time so that the metal can be equilibrated. The hardness and strength obtained depend upon the temperature at which tempering is carried out. Higher temperatures will result into high ductility, but low strength and hardness. Low tempering temperatures will produce low ductility, but high strength and hardness. In practice, appropriate tempering temperatures are selected that will produce the desired level of hardness and strength. This operation is performed on all carbon steels that have been hardened, in order to reduce their brittleness, so that they can be used effectively in desired applications.

Annealing

Annealing involves treating steel up to a high temperature, and then cooling it very slowly to room temperature, so that the resulting microstructure will possess high ductility and toughness, but low hardness. Annealing is performed by heating a component to the appropriate temperature, soaking it at that temperature, and then shutting off the furnace while the piece is in it. Steel is annealed before being processed by cold forming, to reduce the requirements of load and energy, and to enable the metal to undergo large strains without failure.

Normalizing

Normalizing involves heating steel, and then keeping it at that temperature for a period of time, and then cooling it in air. The resulting microstructure is a mixture of ferrite and cementite which has a higher strength and hardness, but lower ductility. Normalizing is performed on structures and structural components that will be subjected to machining, because it improves the machinability of carbon steels.

Carburization

Carburization is a heat treatment process in which steel or iron is heated to a temperature, below the melting point, in the presence of a liquid, solid, or gaseous material which decomposes so as to release carbon when heated to the temperature used.

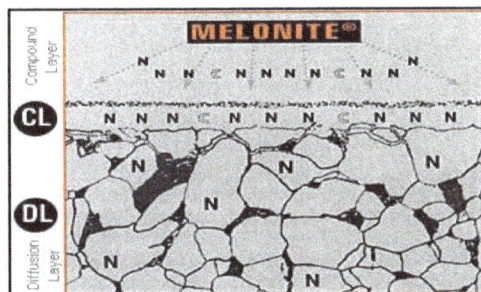

Melonite formation by heat treatment of steels.

The outer case or surface will have higher carbon content than the primary material. When the steel or iron is rapidly cooled by quenching, the higher carbon content on the outer surface becomes hard, while the core remains tough and soft.

Surface Hardening

In many engineering applications, it is necessary to have the surface of the component hard enough to resist wear and erosion, while maintaining ductility and toughness, to withstand impact and shock loading. This is known as surface hardening. This can be achieved by local austentitizing and quenching, and diffusion of hardening elements like carbon or nitrogen into the surface. Processes involved for this purpose are known as flame hardening, induction hardening, nitriding and carbonitriding.

Classification of Steels by Carbon Content

There are only three phases in steels, but there are many different structures. A precise definition of eutectoid carbon is unavoidable; it varies from 0.77% to slightly over 0.80%, depending on the reference used. However, for the objectives of this book, the precise amount of carbon denoted as eutectoid is of no particular significance.

Hypoeutectoid Steels: Carbon steels containing less than 0.80% C are known as hypoeutectoid steels. The area bounded by AGB on the ironcementite diagram is of significance to the room temperature microstructures of these steels; within the area, ferrite and austenite each having different carbon contents, can exist simultaneously.

Assume that a 0.40% carbon steel has been slowly heated until its temperature throughout the piece is $870\,^{\circ}$C ($1600\,^{\circ}$F), thereby ensuring a fully austenitic structure. Upon slow cooling, free ferrite begins to form from the austenite when the temperature drops across the line AG, into the area AGB, with increasing amounts of ferrite forming as the temperature continues to decline while in this area. Ideally, under very slow cooling conditions, all of the free ferrite separates from austenite by the time the temperature of the steel reaches A1 (the line BG) at $725\,^{\circ}$C ($1340\,^{\circ}$F). The austenite islands, which remain at about $725\,^{\circ}$C ($1340\,^{\circ}$F), now have the same amount of carbon as the eutectoid steel, or about 0.80%. At or slightly below $725\,^{\circ}$C ($1340\,^{\circ}$F) the remaining untransformed austenite transforms—it becomes pearlite, which is so named because of its resemblance to mother of pearl. Upon further cooling to room temperature, the microstructure remains unchanged, resulting in a final room temperature microstructure of an intimate mixture of free ferrite grains and pearlite islands.

A typical microstructure of a 0.40% carbon steel is shown in figure. The pure white areas are the islands of free ferrite grains described previously. Grains that are white but contain dark platelets are typical lamellar pearlite. These platelets are cementite or carbide interspersed through the ferrite, thus conforming to the typical two-phase structure indicated below the BH line in figure.

Eutectoid Steels: A carbon steel containing approximately 0.77% C becomes a solid solution at any temperature in the austenite temperature range, i.e., between 725 and $1370\,^{\circ}$C (1340 and $2500\,^{\circ}$F). All of the carbon is dissolved in the austenite. When this solid solution is slowly cooled, several changes occur at $725\,^{\circ}$C ($1340\,^{\circ}$F). This temperature is a transformation temperature or critical temperature of the iron-cementite system. At this temperature, a 0.77% (0.80%) carbon steels transforms from a single homogeneous solid solution into two distinct new solid phases. This change occurs at constant temperature and with the evolution of heat. The new phases are ferrite an cementite, formed simultaneously; however, it is only at composition point G in figure. (0.77% carbon steel) that this phenomenon of the simultaneous formation of ferrite and cementite can occur.

The microstructure of a typical eutectoid steel is shown in figure. The white matrix is alpha ferrite and the dark platelets are cementite. All grains are pearlite—no free ferrite grains are present under these conditions.

Cooling conditions (rate and temperature) govern the final condition of the particles of cementite that precipitate from the austenite at $725\,^\circ$C ($1340\,^\circ$F). Under specific cooling conditions, the particles become spheroidal instead of elongated platelets as shown in figure. Figure shows a similar two-phase structure resulting from slowly cooling a eutectoid carbon steel just below A_1. This structure is commonly known as spheroidite but is still a despersion of cementite particles in alpha ferrite. There is no indication of grain boundaries in figure. The spheroidized structure is often preferred over the pearlitic structure because spheroidite has superior machinability and formability. Combination structures (that is, partly lamellar and partly spheroidal cementite in a ferrite matrix) are also common.

As noted previously, a eutectoid steel theoretically contains a precise amount of carbon. In practice, steels that contain carbon within the range of approximately 0.75 to 0.85% are commonly referred to as eutectoid carbon steels.

Hypereutectoid steels contain carbon contents of approximately 0.80 to 2.0%. Assume that a steel containing 1.0% C has been heated to $845\,^\circ$C ($1550\,^\circ$F), thereby ensuring a 100% austenitic structure. When cooled, no change occurs until the line GF, known as Acm or cementite solubility line, is reached. At this point, cementite begins to separate out from the austenite, and increasing amounts of cementite separate out as the temperature of the 1% carbon steels descends below the A line. The composition of austenite changes from 1% C toward 0.77% C. At a temperature slightly below $725\ ^\circ$C ($1340\,^\circ$F), the remaining austenite changes to pearline. No further changes occur as cooling proceeds toward room temperature, so that the room temperature microstructure consists of pearline and free cementite. In this case, the free cementite exists as a network around the pearline grains.

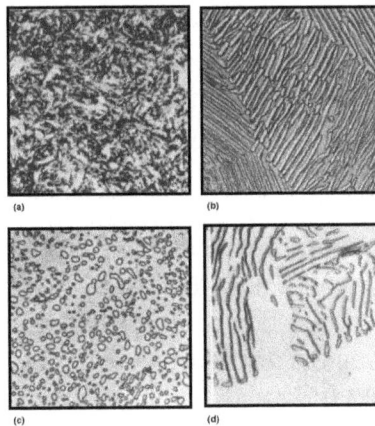

In figure, Effects of carbon content on the microstructures of plain-carbon steels. (a) Ferrite grains (white) and pearlite (gray streaks) in a white matrix of a hypoeutectoid steel containing 0.4% C. 1000×. (b) Microstructure (all pearlite grains) of a eutectoid steel containing 0.77% C. 2000×. (c)

Microstructure of a eutectoid steel containing 0.77% C with all cementite in the spheroidal form. 1000×. (d) Microstructure of a hypereutectoid steel containing ~ 1.0% C containing pearlite with excess cementite bounding the grains. 1000×.

Upon heating hypereutectoid steels, reverse changes occur. At 725 °C (1340 °F), pearlite changes to austenite. As the temperature increases above 725 °C (1340 °F), free cementite dissolves in the austenite, so that when the temperature reaches the A$_{cm}$ line, all the cementite dissolves to form 100% austenite.

Carbon Steel

A carbon steel is an interstitial alloy of only iron and carbon. The carbon has a limited solid state solubility in the iron, and above 25 atomic % carbon a compound of iron and carbon, Fe$_3$C is formed.

The diagram shows the iron-rich end of the iron-carbon phase diagram. The verttical line at 6.7 weight % carbon is Fe$_3$C, and for the field shown this material and iron behave as a binary system. The equilibrium phases are shown on the diagram. Ferrite (α) is bcc iron with carbon in interstitial solid solution. Austenite (γ) is fcc iron with interstitial carbon, and d-iron is bcc iron with interstitial carbon in the high temperature range. Boundaries with two phase zones represents the solubility limit for carbon in these phases.

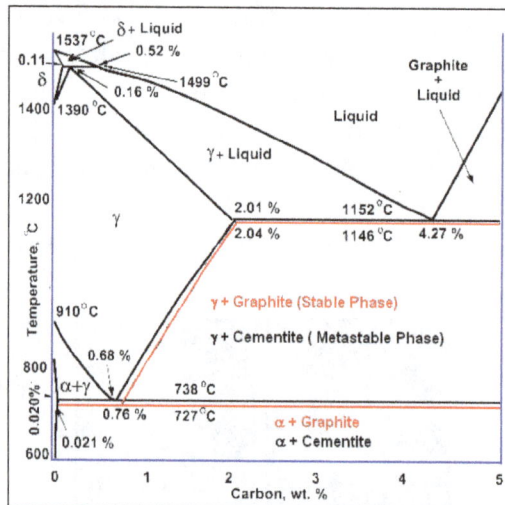

Carbon steel materials are widely used in the oil and gas production industry because of its availability, constructability, and relatively low cost. However, there are limits to the longevity of carbon steelbecause of its low corrosion resistance. In particular, carbon steel tubing and piping are susceptible to erosion–corrosion damage due to the erosive and corrosive nature of the produced fluid or gas. The combined effect of sand erosion and corrosion can be very significant. One form of erosion–corrosion of carbon steels occurs when entrained sand particles impinge the wall and remove part or all of a protective iron carbonate (FeCO$_3$) scale allowing corrosion rates to increase to bare metal rates. This limitation raises the need for reliable prediction tools to properly design production facilities in terms of cost savings and safety.

Carbon Steel Advantages

There are several advantages to choosing carbon steel over traditional steel, one of which is increased strength. The use of carbon makes iron — or steel — stronger by shuffling around its crystal latice. While carbon steel can still stress and break under pressure, it's less likely to occur than with other types of steel. This makes carbon steel particularly effective in applications where strength is needed. Japanese bladesmiths, for example, produced swords out of high-carbon steel known as tamahagane steel many centuries ago. Today, carbon steel is used to make everything from construction materials to tools, automotive components and more.

Carbon Steel Disadvantages

But there are also some disadvantages to choosing carbon steel over traditional steel. Because it's so strong, carbon steel is difficult to work with. It can't be easily bent and molded into different shapes, thus limiting its utility in certain applications. Carbon steel is also more susceptible to rust and corrosion than other types to steel. To make steel "stainless," manufacturers add chromium — usually about 10% to 12%. Chromium acts as a barrier of protection over the steel itself, thereby protecting it from moisture that could otherwise cause rusting. Carbon steel doesn't contain chromium, however, so it may rust when exposed to moisture for long periods of time.

Heat Treatment

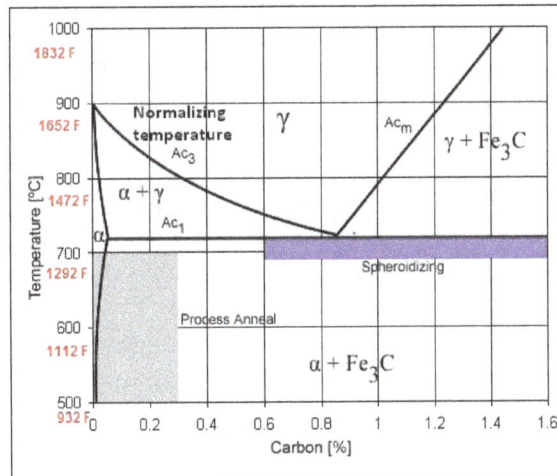

Iron-carbon phase diagram, showing the temperature and carbon ranges for certain types of heat treatments.

The purpose of heat treating carbon steel is to change the mechanical properties of steel, usually ductility, hardness, yield strength, or impact resistance. Note that the electrical and thermal conductivity are only slightly altered. As with most strengthening techniques for steel, Young's modulus (elasticity) is unaffected. All treatments of steel trade ductility for increased strength and vice versa. Iron has a higher solubility for carbon in the austenite phase; therefore all heat treatments, except spheroidizing and process annealing, start by heating the steel to a temperature at which the austenitic phase can exist. The steel is then quenched (heat drawn out) at a moderate to low rate allowing carbon to diffuse out of the austenite forming iron-carbide (cementite) and leaving ferrite, or at a high rate, trapping the carbon within the iron thus forming martensite. The rate at which the steel is cooled through the eutectoid temperature (about 727 °C) affects the rate at which carbon diffuses out of austenite and forms cementite. Generally speaking, cooling swiftly will leave iron carbide finely dispersed and produce a fine grained pearlite and cooling slowly will give a coarser pearlite. Cooling a hypoeutectoid steel (less than 0.77 wt% C) results in a lamellar-pearlitic structure of iron carbide layers with α-ferrite (nearly pure iron) between. If it is hypereutectoid steel (more than 0.77 wt% C) then the structure is full pearlite with small grains (larger than the pearlite lamella) of cementite formed on the grain boundaries. A eutectoid steel (0.77% carbon) will have a pearlite structure throughout the grains with no cementite at the boundaries. The relative amounts of constituents are found using the lever rule. The following is a list of the types of heat treatments possible:

- Spheroidizing: Spheroidite forms when carbon steel is heated to approximately 700 °C for over 30 hours. Spheroidite can form at lower temperatures but the time needed drastically increases, as this is a diffusion-controlled process. The result is a structure of rods or spheres of cementite within primary structure (ferrite or pearlite, depending on which side of the eutectoid you are on). The purpose is to soften higher carbon steels and allow more formability. This is the softest and most ductile form of steel. The adjacent image shows where spheroidizing usually occurs.

- Full annealing: Carbon steel is heated to approximately 40 °C above Ac3? or Acm? for 1 hour; this ensures all the ferrite transforms into austenite (although cementite might still exist if the carbon content is greater than the eutectoid). The steel must then be cooled slowly, in the realm of 20 °C (36 °F) per hour. Usually it is just furnace cooled, where the furnace is turned off with the steel still inside. This results in a coarse pearlitic structure, which means the "bands" of pearlite are thick. Fully annealed steel is soft and ductile, with no internal stresses, which is often necessary for cost-effective forming. Only spheroidized steel is softer and more ductile.

- Process annealing: A process used to relieve stress in a cold-worked carbon steel with less than 0.3% C. The steel is usually heated to 550–650 °C for 1 hour, but sometimes temperatures as high as 700 °C. The image rightward shows the area where process annealing occurs.

- Isothermal annealing: It is a process in which hypoeutectoid steel is heated above the upper critical temperature. This temperature is maintained for a time and then reduced to below the lower critical temperature and is again maintained. It is then cooled to room temperature. This method eliminates any temperature gradient.

- Normalizing: Carbon steel is heated to approximately 55 °C above Ac3 or Acm for 1 hour; this ensures the steel completely transforms to austenite. The steel is then air-cooled, which is a cooling rate of approximately 38 °C (100 °F) per minute. This results in a fine pearlitic structure, and a more-uniform structure. Normalized steel has a higher strength than annealed steel; it has a relatively high strength and hardness.

- Quenching: Carbon steel with at least 0.4 wt% C is heated to normalizing temperatures and then rapidly cooled (quenched) in water, brine, or oil to the critical temperature. The critical temperature is dependent on the carbon content, but as a general rule is lower as the carbon content increases. This results in a martensitic structure; a form of steel that possesses a super-saturated carbon content in a deformed body-centered cubic (BCC) crystalline structure, properly termed body-centered tetragonal (BCT), with much internal stress. Thus quenched steel is extremely hard but brittle, usually too brittle for practical purposes. These internal stresses may cause stress cracks on the surface. Quenched steel is approximately three times harder (four with more carbon) than normalized steel.

- Martempering (Marquenching): Martempering is not actually a tempering procedure, hence the term "marquenching". It is a form of isothermal heat treatment applied after an initial quench, typically in a molten salt bath, at a temperature just above the "martensite start temperature". At this temperature, residual stresses within the material are relieved and some bainite may be formed from the retained austenite which did not have time to transform into anything else. In industry, this is a process used to control the ductility and hardness of a material. With longer marquenching, the ductility increases with a minimal loss in strength; the steel is held in this solution until the inner and outer temperatures of the part equalize. Then the steel is cooled at a moderate speed to keep the temperature gradient minimal. Not only does this process reduce internal stresses and stress cracks, but it also increases the impact resistance.

- Tempering: This is the most common heat treatment encountered, because the final properties can be precisely determined by the temperature and time of the tempering. Tempering involves reheating quenched steel to a temperature below the eutectoid temperature then cooling. The elevated temperature allows very small amounts of spheroidite to form, which restores ductility, but reduces hardness. Actual temperatures and times are carefully chosen for each composition.

- Austempering: The austempering process is the same as martempering, except the quench is interrupted and the steel is held in the molten salt bath at temperatures between 205 °C and 540 °C, and then cooled at a moderate rate. The resulting steel, called bainite, produces an acicular microstructure in the steel that has great strength (but less than martensite), greater ductility, higher impact resistance, and less distortion than martensite steel. The disadvantage of austempering is it can be used only on a few steels, and it requires a special salt bath.

Case Hardening

Case hardening processes harden only the exterior of the steel part, creating a hard, wear resistant skin (the "case") but preserving a tough and ductile interior. Carbon steels are not very hardenable meaning they can not be hardened throughout thick sections. Alloy steels have a better hardenability, so they can be through-hardened and do not require case hardening. This property of carbon steel can be beneficial, because it gives the surface good wear characteristics but leaves the core tough.

Types of Carbon Steel

Low Carbon Steel

Low carbon steel is a type of steel that has small carbon content, typically in the range of 0.05% to 0.3%. Its reduced carbon content makes it more malleable and ductile than other steel types. Low carbon steel is also known as mild steel.

Low carbon steel is one of the most common types of steel. Low carbon steel is ideal for applications in which precision is paramount due to its heightened flexibility. It is less prone to corrosion than other types of steel due to its reduced carbon content.

Properties of Low Carbon Steel

In order for steel to be considered low carbon steel, there are certain characteristics it must meet. For instance, the steel has to have less than .3 percent carbon in its total makeup to be considered low carbon. Low carbon steel also contains pearlite and ferrite as major components. Low carbon steel is generally used straight from the forming process, whether that process is hot forming or cool forming, because that's when it's most workable and easiest to form.

Weldability

Low carbon steel has some of the best weldability of any metal. The reason for this is precisely due to the low carbon content of the metal. As carbon is added to steel, the steel gets harder and harder. This is a desirable outcome if the steel is going to be used structurally, or in a situation where strength is of the utmost importance. However, the harder the steel gets with more carbon, the more prone to cracking it is when you attempt to weld it. As such, low carbon steel doesn't have that problem.

Formability

Low carbon steel also possesses good formability. This means that low carbon steel is easier to form into certain shapes, through such methods as pouring, molding and pressing. Also, low carbon steel is used for case hardened machine parts, chain, rivets, stampings, nails, wire and pipe. The ability of the steel to be turned into a number of different forms makes it quite versatile. When using low carbon steel, strength isn't the primary concern, because what you lose in rigidity, you gain in formability.

Medium Carbon Steel

Medium-carbon steels are similar to low-carbon steels except that they contain carbon from 0.30% to 0.60% and manganese from 0.60% to 1.65%. Increasing the carbon content to approximately 0.5% with an accompanying increase in manganese allows medium-carbon steels to be used in the quenched and tempered condition. These steels are mainly used for making shafts, axles, gears, crankshafts, couplings, and forgings. Steels with carbon ranging from 0.40% to 0.60% are used for rails, railway wheels, and rail axles.

Uses for Medium-Carbon Steel

The uses for medium-carbon steel are defined by the requirement for a high tensile strength and ductility that, despite its brittleness when compared to other forms of steel, make it the preferred choice. Between 0.3 and 0.7 percent carbon is added during the manufacturing process to create a medium or mid-range steel product. This specific range of carbon is combined with a process of quenching (i.e., cooling the steel from the outer surface to the inner) and tempering to create a structure that has a consistent tensile strength (referred to as Martensite) throughout the body.

Medium-carbon steel is used in high-tension applications.

Shafts and Gearing

Axle shafts, crankshafts and gearing plates are all made from medium-carbon steel. The ductility of the steel allows it to be formed into thin shafts or toothed plates without losing any of its tensile strength.

Pressured Structures

The ductility of the medium-carbon steel allows it to be shaped into plates for boilers and other tanks that have highly pressurized contents. Medium-carbon steel cannot be used for pressurized tank systems that contain cold liquids or gasses because the Martensite structure of the steel makes it brittle and susceptible to cold cracking. Stainless steel or other high carbon steels are used for those types of applications.

Railway Applications

Railway wheels, rails and other steel parts associated with the suspension of rail cars are made of medium-carbon steel. The high tensile strength is necessary to withstand the changing force of the rail cars on the rails.

Structural Steel

Structural steel beams, joiner plates and other shapes associated with building require a high tensile strength to resist the torque and pressure of buildings and bridges. Special care must be taken to properly insulate the steel to prevent it from being affected by extremes of heat and cold, which can change the Martensite structure and lessen its structural integrity.

High Carbon Steel

High carbon steels are steels with high carbon content. If iron is heated to a high temperature, it dissolves carbon, which would normally precipitate upon cooling. However, if this liquid metal is cooled very quickly by 'quenching' it in water, the carbon is trapped and distorts the structure of the substance, forming high carbon steel. If you continue increasing the carbon content of steel beyond about 2%, it eventually becomes cast iron, the material famously used in Le Creuset cookware, a much harder, denser and more brittle metal as a result of its impurities. The carbon in this sample of steel strengthens it and gives it the ability to harden by heat treatment. It also makes it less ductile and weldable than ordinary steel, and it becomes much more brittle as a result of its impurities. Because of its extreme hardness and resistance to wear, high carbon steel is often used for things like cutting tools that retain their sharp edge, and masonry nails that can be driven into concrete blocks or bricks without bending (although due to their brittleness they do have a propensity to fracture if they are mistreated). As you can see from the flecks of rust on the surface of this sample and its blackened sides, unlike stainless steel, this metal is not resistant to oxidation. However, the carbon content of this sample does mean it corrodes much more slowly than low carbon steel or iron.

High carbon steel has a reputation for being especially hard, but the extra carbon also makes it more brittle than other types of steel. This type of steel is the most likely to fracture under stress.

Hardness and other Advantages

High carbon steel has important advantages over other materials. This type of steel is excellent for making cutting tools or masonry nails. The carbon gives the steel hardness and strength while being relatively inexpensive compared to other hard substances. Manufacturers value high carbon steel for metal cutting tools or press machinery that bends and forms metal parts.

Brittleness and other Disadvantages

Some disadvantages also come with the use of high carbon steel. It is difficult to weld, posing challenges for manufacturers and fabricators. The same quality of hardness that makes it preferred for cutting tools also means it is brittle, making it prone to fracture or break. It also doesn't hold up to wear as well as other types of specialty steel. Tools made with high carbon steels can become magnetized over time, attracting unwanted iron dust and particles.

Medium and high carbon steels are widely used in many common applications. Increasing carbon as the primary alloy for the higher strength and hardness of steels is usually the most economical approach to improved performance. However, some of the effects of elevated carbon levels include reduced weldability, ductility and impact toughness. When these reduced properties can be tolerated, the increased strength and hardness of the higher carbon materials can be used to a significant advantage. Common applications of higher carbon steels include forging grades, rail steels, spring steels (both flat rolled and round), pre-stressed concrete, wire rope, tire reinforcement, wear resistant steels (plates and forgings), and high strength bars.

To increase the performance of steels in these applications, it is common to maximize strength and hardness by raising the carbon level to the highest practical level. The limiting factor to carbon additions will vary depending on the type of applications. For forging steels and bar products, it may be toughness or weldability. For high strength wire, the limiting factor for carbon addition is generally the eutectoid carbon level, above which the presence of grain boundary carbides will dramatically reduce drawability.

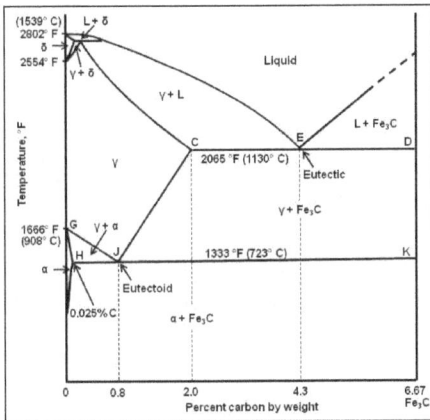

Historical Description of the Fe-C phase diagram.   Microstructure of high carbon steel.

Generally, the high carbon steels contain from 0.60 to 1.00% C with manganese contents ranging from 0.30 to 0.90%. High carbon steels are used for spring materials and high-strength wires. Ultrahigh carbon steels are experimental alloys containing approximately 1.25 to 2.0% C. These steels are thermomechanically processed to produce microstructures that consist of ultrafine, equiaxed grains of ferrite and a uniform distribution of fine, spherical, discontinuous proeutectoid carbide particles. Such microstructures in these steels have led to superplastic behavior.

Figure shows the microstructure of high carbon steel with about 0.8% C by weight, alloyed with iron. The steel has one major constituent, which is pearlite. It is made up from a fine mixture of ferrite and iron carbide, which can be seen as a "wormy" texture. The pearlite has a very fine structure, which makes the steel very hard. Unfortunately this also makes the steel quite brittle and much less ductile than mild steel. The high carbon steel has good wear resistance, and until recently was used for railways. It is also used for cutting tools, such as chisels and high strength wires. These applications require a much finer microstructure, which improves the toughness.

Alloy Steel

Alloy steel is a type of steel that has undergone alloying using different elements in levels between 1% and 50% in weight in order to enhance mechanical properties. It can be classified further into two types: high-alloy and low-alloy steels.

Alloy steels possess properties like increased durability and higher resistance to corrosion.

However, alloy steel is any steel where one or more of its elements aside from carbon have been added intentionally to achieve more desirable characteristics.

Some of the most common elements that are added to generate alloy steels include:

- Manganese
- Silicon
- Molybdenum
- Chromium
- Vanadium
- Nickel

The difference is somewhat uniform, but to make it distinguishable, all steel alloyed with higher than 8% of its weight of elements other than carbon and alloy is considered high-alloy steel. Alloyed steels are harder, more durable and more resistant to corrosion.

Alloy steels with carbon levels of medium to elevated rates are difficult to weld. However, if the carbon levels are reduced to 1% to 3%, such alloy metals can achieve greater formability and weldability, thus, improved strength.

When other elements comprising metals and non-metals are added to carbon steel, alloy steel is formed. The composition and proportion of alloying elements determine the various properties of alloy steel.

Steel

Steel is among the most popular materials used in the construction industry. According to the World Steel Association, in 2018 around 1,808 million tons of crude steel was produced worldwide and about 50% of this production was utilized by the construction industry. Further, they also state that there are as many as 3,500 different grades of steel and each grade offers environmental, chemical and physical properties unique to that grade of steel. Steel has undergone significant evolution through time and around 75% of all the types of modern-day steel were developed in the past 20 years. It is interesting to note that had the Eiffel Tower (constructed in 1887) been constructed in today's times, it would require only one-third of the steel used back then.

Types of Steel

Fundamentally, steel is an alloy of iron with low amounts of carbon. There are thousands of different types of steels which are created to suit different kinds of applications. These broadly fall into 4 types – carbon steel, tool steel, stainless steel and alloy steel. Carbon steels form the majority of steels produced in the world today. Tool steels are used to make machine parts, dies and tools. Stainless steels are used to make common household items. Alloy steels are made of iron, carbon and other elements such as vanadium, silicon, nickel, manganese, copper and chromium.

Alloy Steel

When other elements comprising metals and non-metals are added to carbon steel, alloy steel is formed. These alloy steels display various environmental, chemical and physical properties that can vary with the elements used to alloy. Here the proportion of alloying elements can provide different mechanical properties.

Effects of Alloying

Alloying elements can alter carbon steel in several ways. Alloying can affect micro-structures, heat-treatment conditions and mechanical properties. Today's technology with high-speed computers can foresee the properties and micro-structures of steel when it is cold-formed, heat treated, hot-rolled or alloyed. For instance, if properties such as high strength and weldability are required in steel for certain applications, then carbon steel alone will not serve the purpose because carbon's inherent brittleness will make the weld brittle. The solution is to reduce carbon and add other elements such as manganese or nickel. This is one way of making high strength steel with required weldability.

Types of Alloy Steel

There are two kinds of alloy steel – low-alloy steel and high-alloy steel. As mentioned earlier, the composition and proportion of alloying elements determine the various properties of alloy steel. Low-alloy steels are the ones which have up to 8% alloying elements whereas high-alloy steels have more than 8% alloying elements.

Alloying Elements

There are around 20 alloying elements that can be added to carbon steel to produce various grades of alloy steel. These provide different types of properties. Some of the elements used and their effects include:

- Aluminium – can rid steel of phosphorous, sulfur and oxygen.
- Chromium – can increase toughness, hardness and wear resistance.
- Copper – can increase corrosion resistance and harness.
- Manganese – can increase high-temperature strength, wear resistance, ductility and hardenability.
- Nickel – can increase corrosion, oxidation resistance and strength.
- Silicon – can increase magnetism and strength.
- Tungsten – can increase strength and hardness.
- Vanadium – can increase corrosion, shock resistance, strength and toughness.

Other alloying elements that provide varied properties include bismuth, cobalt, molybdenum, titanium, selenium, tellurium, lead, boron, sulfur, nitrogen, zirconium and niobium. These alloying elements can be used singly or in various combinations depending on the properties desired.

Alloy Steel Products and their Applications

There are hundreds of products that can be manufactured with alloy steels of varying compositions. These include alloy steel pipes and tubes, alloy steel plates, sheets and coils, alloy steel bars, rods and wires, alloy steel forged fittings, alloy steel buttweld fittings, alloy steel flanges, fasteners and more. Alloy steels have many uses in various industries such as automobiles, mining, machinery and equipment, railways, road construction, buildings, appliances and off-shore applications.

Applications in Building Large Structures

In the building and construction industry, alloy steels are used for very large modern structures such as airports, bridges, skyscrapers and stadiums in the form of a steel skeleton. Alloy steels provide the required high strength to support such large structures. Even concrete structures use alloy steels as reinforcement to add strength and reduce the overall weight of structures. Smaller items such as screws, nails and bolts made of alloy steels are used in the building and construction industry.

Applications in Building Bridges

Bridges use special alloy steels known as weathering steels. These provide enhanced protection from corrosion because of the presence of nickel, copper and chromium as alloying elements. Weathering steels also find uses in buildings as facing material to improve aesthetics. Weathering steel offers several benefits which include high safety, ease and quickness of construction, aesthetic looks, shallow depth of construction, low maintenance and amenable to alterations in the future. Because of its natural weathered finish, no painting is required thus avoiding environmental issues caused by paints. Weathering steels are extremely cost-effective in the long run.

Alloy Steel Flat Products

Alloy steels are used to make flat products – plates and strips. Plates are available in a wide range of grades and sizes. These are used in building construction by welding plates into fabricated sections.

Alloy Steel Strip and Coil Products

Strips are available as hot and cold rolled strips and hot-dipped galvanized coils. Hot-dipped galvanized coils are used to make construction products that include wall and roof claddings, side rails, roof purlins, light steel frames and lintels.

Alloy Steel Long Products

Alloy steels are employed in producing long products used in the construction industry such as girders, structural sections, bars, rails, rods and wires.

Alloy Steel Flanges

Another important product made of alloy steels are flanges. These are used in stainless steel pipelines. These flanges can be made to suit various applications. Some of them include weld neck flanges which have the same thickness and bevel as of the pipe and can perform well under harsh conditions of high pressure, high temperature or sub-zero temperatures. Lap joint flanges are slip-on flanges suitable for alloy steel piping which require regular maintenance and inspection.

Alloy Steel Pipe Products

Alloy steel pipes are important materials in the building and construction industry because of their performance characteristics such as ductility, ease of fixing without heat treatment and high durability. They are an alloy of stainless steel, chromium and nickel. Some special types of alloy steel pipes include large diameter welded pipes, electric fusion welded pipes, welded pipes and seamless pipes. They are extremely useful for high-temperature or corrosive applications besides high-pressure environments.

Low-Alloy Steel

Low alloy steel is a sort of steel that has different materials added to it; however alternate materials commonly make up a little measure of the whole steel. Through the expansion of specific alloys, low-alloy steel have exact concoction pieces and give preferred mechanical properties over numerous traditional gentle or carbon steels. Steel ordinarily is an alloy comprising of carbon and iron, however low alloy steel regularly includes hard metals, for example, nickel and chromium. These alloys normally include one to five percent of the steel's substance and are included based their capacity to give a particular trait. This steel, for the most part, is hard and hard to weld, yet the steel's qualities can be changed relying upon the incorporations. Luckily, in spite of the expansion of these components, low-alloy steel aren't really hard to weld. This most regularly is utilized to make huge funnels for the oil business; however, it additionally can be utilized for development and military purposes. In any case, knowing precisely what sort of low-alloy steel you have is basic to accomplishing great weld honesty, as is legitimate filler metal choice. Steel itself is an exceptionally essential alloy that uses a blend of iron and carbon.

Company Products Compatible With Low Alloy Steels

- High heat resistance steel (Chromium- Molybdenum Steel)
- Low temperature use steel (Nickel Steel)
- Weathering Steel
- High yield and high tensile strength steel

Low Alloy Steels General Description

Chromium-molybdenum Steel

This low alloy steel series contains 0.5% ~ 9% Cr and 0.5% ~ 1% Mo. Its carbon content on average is lower than 0.20%, with decent weldability and higher hardening ability due to its alloy trait. The Cr content improves its anti-oxidization and anti-corrosion ability, and Mo enhances its strength in high temperature condition; The steel supplying conditions are generally gone through annealing or normalizing and tempering processes. Chromium-Molybdenum Steel has been widely used in the areas such as petrol chemical industry, steam power equipment, and high temperature services.

Nickel Steel

The average steel in low temperature environment will have higher strength but lower elongation and toughness, thus increases the chance for brittle fracture. If the steel is needed in a low temperature environment, having superior low temperature toughness is essential. Any suitable steel for this purpose is called low temperature service steel or Nickel steel. Low Alloy Low Temperature Service Steel is formed by adding 2.5% to 3.5 % of Ni in the carbon steel to enhance its low temperature toughness. Ni can strengthen ferrite matrix while lowering Ar3 (third transformation temperature) which helps with fi ne grain formation. In addition to the normalizing treatment during the production process of low alloy low temperature service steel, quenching and tempering are also parts of the mechanical properties improvement treatment.

Weathering Steel

Generally there are two categories of rust prevention methods: one type for instance, is paint coating, electroplate, ceramic coating, or adding layers of anti-corrosion material, anything to shield the steel surface from corrosive environment. Another type is to use stainless steel or weather steel, meaning adding anti-corrosion alloy elements into the steel. Weathering steel is formed by adding small amounts of Cu, Cr, P, Ni, and other alloy elements into low alloy steel. During the initial application, it will also rust like the average carbon steel; however, after certain period (usually one year) the rustic surface will serve as an impermeable protective cover, preventing the further expansion of rust into inner part of steel.

High Tensile and High Yield Strength Steel

This low alloy steel series are added Mn, Ni, Cr, and Mo etc, can increase strength of ferrite matrix; improve the hardening tendency; and allow better control of grain size. This type of steel under as welded condition can meet the demand of high strength, corrosion resistance, or improve notch toughness and other mechanical properties.

This steel type has good weld ability with the yield strength from 70 to 120 ksi, and tensile strength from 90 to 150 ksi.

High Strength Low-alloy Steels (HSLA)

High strength low-alloy steels (HSLA) refers to alloy steel that have higher strength value. Compared to carbon steel, HSLA steels have better corrosion resistance as well as greater mechanical properties. HSLA steels are not defined by their alloying elements or composition. Rather, the improvement of mechanical properties by an alteration of the microstructure determines whether steel can be defined as HSLA.

High strength low-alloy steels provide better mechanical properties compared to carbon steel. Generally speaking, grain size is reduced to reduce pearlite structure, increasing the material's yield strength. Typical elements that are added to achieve this are titanium, copper, niobium and vanadium. Carbon contents of HSLA steels can be anywhere between 0.05 and 0.25% (in mass content) in order to retain formability and weldability.

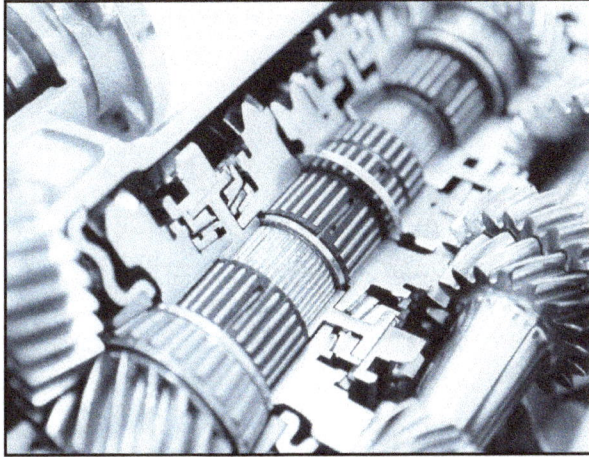

Various alloying elements can be added for different effects other than just strengthening. For example, nitrogen can be added to improve wear protection and resistance to localized corrosion. Other elements include, but are not limited to, nickel, chromium, molybdenum and calcium. However, as low-alloy steels, the combined amount of alloying elements (excluding carbon) does not exceed the limit of 2% in mass content.

Due to its altered microstructure, HSLA material does not rust as fast as carbon steel because of its ferrite structure. That does not mean that HSLA steel can not rust at all. Rust prevention, or the spreading thereof, can be influenced by using certain alloying elements, such as chromium, which forms a protective layer of chromium oxides instead of iron oxide.

Classifications

Microalloyed Steels

Microalloyed steels contain very small amounts of alloying elements (0.05 – 0.15%), meaning they are very low-alloy steel. Without any heat treatment, its yield strength is only 500 to 750 MPa, which is close to the yield strength of carbon steel (415 MPa). Weldability can be improved by reducing carbon contents to 0.05%. Microalloyed steels can be cold-or hot-worked to achieve greater ductility or and mechanical strength. On the plus side, microalloyed steels are not prone to crack due to quenching nor do they have to be straightened.

Weathering Steels

Weathering steels are high strength low alloy steels that are known for their high corrosion and abrasion resistance compared to other steels. The term 'weathering' is derived from the word 'weather' because this type of steel forms a layer on its surface for protection against weather influences. This protective layer develops due to different alloying elements: carbon, silicon, manganese, phosphorus, sulfur, chromium, copper, vanadium, and nickel.

The reason why weathering steel has a high corrosion resistance is not because it does not rust. Quite on the contrary, the steel has to rust to form its protective layer. Carbon steels form iron oxides on their surface which spreads not only on the surface, but can also cause cavitation, meaning that oxidation can penetrate deeper into the material. The alloying elements of weathering steels however, form strong oxides (rust) on the surface, preventing any deeper corrosion. This is also why weathering steels can be recognized by their rusty color.

Pearlite-reduced Steels

Pearlite refers to the atomic structure of materials. It is made up of alternating strips of ferrite (body-centered cubic) and cementite (orthorhombic structure). Perlitic steels are known for their high hardness and high yield strength. To form a fully pearlitic structure, there has to be at least 0.8% carbon content. Due to the high content of carbon, pearlite steels are more susceptible to abrasive wear and cutting force. Pearlite-reduced steels aim to improve these mechanical properties by producing fine grain ferrites. Therefore, there's little to no pearlite in the microstructure.

Dual-phase Steels

Dual-phased steels have a ferritic-martensitic structure. Therefore, they have a high ultimate tensile strength and low initial yielding stress. Dual-phase steels are more malleable than microalloyed steels and show great fatigue resistance. Furthermore, they are deoxidized, meaning all of the oxygen is removed from the material during the steelmaking process, reducing gas porosity greatly. Dual-phase steels are often used for automotive parts such as wheels.

High Alloy Steel

High Alloy Steel is basically an alloy of Iron which consists of Chromium of 10.5%. High alloy steel likewise has over 10% mixture of the alloy. Chromium delivers a thin layer of oxide on the surface of the steel known as the latent layer. They are smidgen costly than low-alloy steel. This keeps any further consumption of the surface. High level of carbon and manganese are added to give austenitic nature to steel. Expanding the measure of Chromium gives an expanded protection from erosion. Due to the high chromium content, high-alloy steel can oppose consumption. High-alloy steel additionally contains shifting measures of Silicon, Manganese, and Carbon. Utilized for benefit in extraordinary hot gasses and fluids and at high temperatures Different components, for example, Molybdenum and Nickel might be added to grant other helpful properties, for example, improved formability and expanded consumption protection.

Stainless Steel

Stainless steel is an iron-containing alloy—a substance made up of two or more chemical elements—used in a wide range of applications. It has excellent resistance to stain or rust due to its chromium content, usually from 12 to 20 percent of the alloy. There are more than 57 stainless steels recognized as standard alloys, in addition to many proprietary alloys produced by different stainless steel producers. These many types of steels are used in an almost endless number of applications and industries: bulk materials handling equipment, building exteriors and roofing, automobile components (exhaust, trim/decorative, engine, chassis, fasteners, tubing for fuel lines), chemical processing plants (scrubbers and heat exchangers), pulp and paper manufacturing, petroleum refining, water supply piping, consumer products, marine and shipbuilding, pollution control, sporting goods (snow skis), and transportation (rail cars), to name just a few.

About 200,000 tons of nickel-containing stainless steel is used each year by the food processing industry in North America. It is used in a variety of food handling, storing, cooking, and serving equipment—from the beginning of the food collection process through to the end. Beverages such as milk, wine, beer, soft drinks and fruit juice are processed in stainless steel equipment. Stainless steel is also used in commercial cookers, pasteurizers, transfer bins, and other specialized equipment. Advantages include easy cleaning, good corrosion resistance, durability, economy, food flavor protection, and sanitary design. According to the U.S. Department of Commerce, 1992 shipments of all stainless steel totaled 1,514,222 tons.

Stainless steels come in several types depending on their microstructure. Austenitic stainless steels contain at least 6 percent nickel and austenite—carbon-containing iron with a face-centered cubic structure—and have good corrosion resistance and high ductility (the ability of the material to bend without breaking). Ferritic stainless steels (ferrite has a body-centered cubic structure) have better resistance to stress corrosion than austenitic, but they are difficult to weld. Martensitic stainless steels contain iron having a needle-like structure.

Duplex stainless steels, which generally contain equal amounts of ferrite and austenite, provide better resistance to pitting and crevice corrosion in most environments. They also have superior resistance to cracking due to chloride stress corrosion, and they are about twice as strong as the common austenitics. Therefore, duplex stainless steels are widely used in the chemical industry in refineries, gas-processing plants, pulp and paper plants, and sea water piping installations.

Raw Materials

Stainless steels are made of some of the basic elements found in the earth: iron ore, chromium, silicon, nickel, carbon, nitrogen, and manganese. Properties of the final alloy are tailored by varying the amounts of these elements. Nitrogen, for instance, improves tensile properties like ductility. It also improves corrosion resistance, which makes it valuable for use in duplex stainless steels.

The Manufacturing Process

The manufacture of stainless steel involves a series of processes. First, the steel is melted, and then it is cast into solid form. After various forming steps, the steel is heat treated and then cleaned and polished to give it the desired finish. Next, it is packaged and sent to manufacturers, who weld and join the steel to produce the desired shapes.

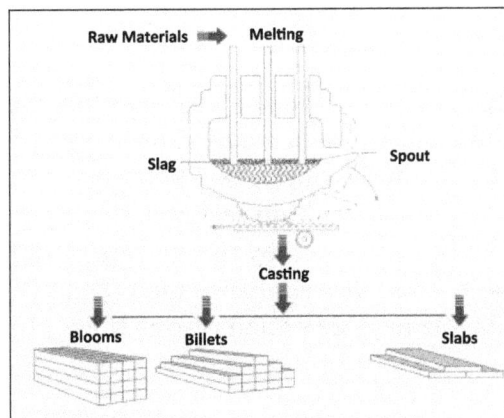

To make stainless steel, the raw materials—iron ore, chromium, silicon, nickel, etc.—are melted together in an electric furnace. This step usually involves 8 to 12 hours of intense heat. Next, the mixture is cast into one of several shapes, including blooms, billets, and slabs.

Melting and Casting

The raw materials are first melted together in an electric furnace. This step usually requires 8 to 12 hours of intense heat. When the melting is finished, the molten steel is cast into semi-finished forms. These include blooms (rectangular shapes), billets (round or square shapes 1.5 inches or 3.8 centimeters in thickness), slabs, rods, and tube rounds.

Forming

Next, the semi-finished steel goes through forming operations, beginning with hot rolling, in which the steel is heated and passed through huge rolls. Blooms and billets are formed into bar and wire, while slabs are formed into plate, strip, and sheet. Bars are available in all grades and come in rounds, squares, octagons, or hexagons 0.25 inch (.63 centimeter) in size. Wire is usually available up to 0.5 inch (1.27 centimeters) in diameter or size. Plate is more than 0.1875 inch (.47 centimeter) thick and over 10 inches (25.4 centimeters) wide. Strip is less than 0.185 inch (.47 centimeter) thick and less than 24 inches (61 centimeters) wide. Sheet is less than 0.1875 (.47 centimeter) thick and more than 24 (61 centimeters) wide.

Heat Treatment

After the stainless steel is formed, most types must go through an annealing step. Annealing is a heat treatment in which the steel is heated and cooled under controlled conditions to relieve internal stresses and soften the metal. Some steels are heat treated for higher strength. However, such a heat treatment—also known as age hardening —requires careful control, for even small changes from the recommended temperature, time, or cooling rate can seriously affect the properties. Lower aging temperatures produce high strength with low fracture toughness, while higher-temperature aging produces a lower strength, tougher material.

Though the heating rate to reach the aging temperature (900 to 1000 degrees Fahrenheit or 482 to 537 degrees Celsius) does not effect the properties, the cooling rate does. A post-aging quenching (rapid cooling) treatment can increase the toughness without a significant loss in strength. One such process involves water quenching the material in a 35-degree Fahrenheit (1.6-degree Celsius) ice-water bath for a minimum of two hours.

The type of heat treatment depends on the type of steel; in other words, whether it is austenitic, ferritic, or martensitic. Austenitic steels are heated to above 1900 degrees Fahrenheit (1037 degrees Celsius) for a time depending on the thickness. Water quenching is used for thick sections, whereas air cooling or air blasting is used for thin sections. If cooled too slowly, carbide precipitation can occur. This buildup can be eliminated by thermal stabilization. In this method, the steel is held for several hours at 1500 to 1600 degrees Fahrenheit (815 to 871 degrees Celsius). Cleaning part surfaces of contaminants before heat treatment is sometimes also necessary to achieve proper heat treatment.

Descaling

Annealing causes a scale or build-up to form on the steel. The scale can be removed using several processes. One of the most common methods, pickling, uses a nitric-hydrofluoric acid bath to descale the steel. In another method, electrocleaning, an electric current is applied to the surface using a cathode and phosphoric acid, and the scale is removed. The annealing and descaling steps occur at different stages depending on the type of steel being worked. Bar and wire, for instance, go through further forming steps (more hot rolling, forging, or extruding) after the initial hot rolling before being annealed and descaled. Sheet and strip, on the other hand, go through an initial annealing and descaling step immediately after hot rolling. After cold rolling (passing through rolls at a relatively low temperature), which produces a further reduction in thickness, sheet and strip are annealed and descaled again. A final cold rolling step then prepares the steel for final processing.

Cutting

Cutting operations are usually necessary to obtain the desired blank shape or size to trim the part to final size. Mechanical cutting is accomplished by a variety of methods, including straight shearing using guillotine knives, circle shearing using circular knives horizontally and vertically positioned, sawing using high speed steel blades, blanking, and nibbling. Blanking uses metal punches and dies to punch out the shape by shearing. Nibbling is a process of cutting by blanking out a series of overlapping holes and is ideally suited for irregular shapes.

Stainless steel can also be cut using flame cutting, which involves a flame-fired torch using oxygen and propane in conjunction with iron powder. This method is clean and fast. Another cutting method is known as plasma jet cutting, in which an ionized gas column in conjunction with an electric arc through a small orifice makes the cut. The gas produces extremely high temperatures to melt the metal.

Finishing

Surface finish is an important specification for stainless steel products and is critical in applications where appearance is also important. Certain surface finishes also make stainless steel easier to clean, which is obviously important for sanitary applications. A smooth surface as obtained by polishing also provides better corrosion resistance. On the other hand, rough finishes are often required for lubrication applications, as well as to facilitate further manufacturing steps.

Surface finishes are the result of processes used in fabricating the various forms or are the result of further processing. There are a variety of methods used for finishing. A dull finish is produced by hot rolling, annealing, and descaling. A bright finish is obtained by first hot rolling and then cold rolling on polished rolls. A highly reflective finish is produced by cold rolling in combination with annealing in a controlled atmosphere furnace, by grinding with abrasives, or by buffing a finely ground surface. A mirror finish is produced by polishing with progressively finer abrasives, followed by extensive buffing. For grinding or polishing, grinding wheels or abrasive belts are normally used. Buffing uses cloth wheels in combination with cutting compounds containing very fine abrasive particles in bar or stick forms. Other finishing methods include tumbling, which forces movement of a tumbling material against surfaces of parts, dry etching (sandblasting), wet etching using acid solutions, and surface dulling. The latter uses sandblasting, wire brushing, or pickling techniques.

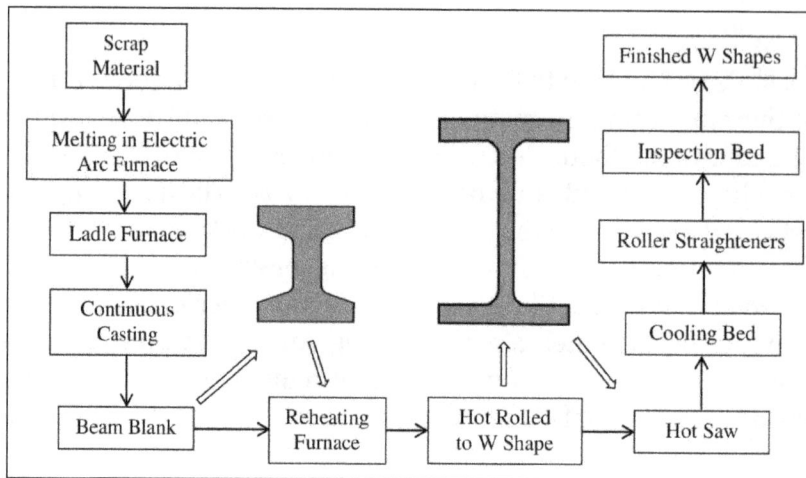

The initial steel shapes—blooms, billets, slabs, etc.—are hot rolled into bar, wire, sheet, strip, and plate. Depending on the form, the steel then undergoes further rolling steps (both hot and cold rolling), heat treatment (annealing), descaling Ito remove buildup), and polishing to produce the finished stainless steel. The steel is then sent the end user.

Manufacturing at the Fabricator or end user

After the stainless steel in its various forms are packed and shipped to the fabricator or end user, a variety of other processes are needed. Further shaping is accomplished using a variety of methods, such as roll forming, press forming, forging, press drawing, and extrusion. Additional heat treating (annealing), machining, and cleaning processes are also often required.

There are a variety of methods for joining stainless steel, with welding being the most common. Fusion and resistance welding are the two basic methods generally used with many variations for both. In fusion welding, heat is provided by an electric arc struck between an electrode and the metal to be welded. In resistance welding, bonding is the result of heat and pressure. Heat is produced by the resistance to the flow of electric current through the parts to be welded, and pressure is applied by the electrodes. After parts are welded together, they must be cleaned around the joined area.

Stainless Steels Alloying Elements

Carbon

Carbon and iron are alloyed together to form steel. This process boosts the strength and hardness of iron. Heat treatment is not adequate to strengthen and harden pure iron, but when carbon is added, a wide range of strength and hardness is realized.

High carbon content is not preferred in Ferritic and Austenitic stainless steels, specifically for welding purposes, due to the risk of carbide precipitation.

Manganese

The addition of manganese to steel improves hot working properties and boosts toughness, strength, and hardenability. Just like nickel, manganese is an Austenite forming element and has been traditionally used as a replacement for nickel in the AISI200 range of Austenitic stainless steels, for example AISI 202 as a replacement for AISI 304.

Chromium

Chromium is combined with steel to improve it's resistance to oxidation. When more chromium is added, the resistance is improved further.

Stainless steels have at least 10.5% chromium (usually 11 or 12%), which imparts a considerable level of corrosion resistance, compared to steels with a relatively lower percentage of chromium.

The resistance to corrosion is attributed to the formation of a passive, self-repairing layer of chromium oxide on the stainless steel surface.

Nickel

Large amounts of nickel - more than 8% - is added to high chromium stainless steels to produce the most important group of steels that are resistant to both heat and corrosion.

These include the Austenitic stainless steels that are characterized by 18-8 (304/1.4301), where nickel's tendency to form Austenite contributes to high strength and excellent toughness or impact strength, at both low and high temperatures. Nickel also significantly improves resistance to corrosion and oxidation.

Molybdenum

When mixed with chromium-nickel austenitic steels, molybdenum enhances resistance to crevice and pitting corrosion, particularly in sulphur and chlorides-containing environments.

Nitrogen

Similar to nickel, nitrogen is an Austenite forming element and increases the Austenite stability of stainless steels. When nitrogen is mixed with stainless steels, yield strength is considerably enhanced along with increased resistance to pitting corrosion.

Copper

In stainless steel, copper is often present as a residual element. This element is added to several alloys to create precipitation hardening characteristics or to improve corrosion resistance, predominantly in sulphuric acid and sea water conditions.

Titanium

Titanium is often added to stabilize carbide, particularly when the material has to be welded. Titanium merges with carbon to form titanium carbides that are relatively stable and cannot be easily dissolved in steel, which is likely to reduce the occurrence of inter-granular corrosion.

When around 0.25 / 0.60% titanium is added, it causes the carbon to merge with titanium as opposed to chromium, avoiding a tie-up of corrosion-resistant chromium as inter-granular carbides and the associated loss of corrosion resistance at the grain boundaries.

In the past several years, the use of titanium has considerably reduced because of the ability of steelmakers to supply stainless steels that have extremely low carbon contents. Such steels can be readily welded without any need for stabilization.

Phosphorus

In order to improve machinability, phosphorus is often added with sulphur. While the presence of phosphorus in Austenitic stainless steels boosts strength, it has a detrimental effect on corrosion resistance and increases the material's tendency to break during welding.

Sulphur

Sulphur improves machinability when it is added in small quantities, but just like phosphorous, it has a negative effect on corrosion resistance and the subsequent weldability.

Selenium

Selenium was previously employed as an addition to enhance machinability.

Niobium/Colombium

Carbon stabilization is achieved by adding niobium to steel, and performs in the same manner as titanium. In addition, niobium strengthens alloys and steels for increased temperature service.

Silicon

Silicon is typically employed as a deoxidizing (killing) agent in the steel melting process, and a small amount of silicon is used in most steels.

Cobalt

When subjected to strong radiation of nuclear reactors, cobalt becomes highly radioactive and hence, all stainless steels deployed in nuclear service will have certain cobalt limitation, often 0.2% at the most.

This issue is important as some amount of the remaining cobalt will be present in the nickel used to make Austenitic stainless steels.

Calcium

Calcium is added in small amounts to enhance machiniability, without having any detrimental effect on other properties induced by selenium, phosphorus and sulphur.

Types of Stainless Steel

Austenitic Stainless Steel

Austenitic stainless steel is the type of stainless steel that is used most commonly in industry for up to 70% of all the stainless steel production because it can be formed and welded easily with successful results. One of the reasons for using the austenitic stainless steel is that this steel can withstand cryogenic temperatures (-238 °F, -150 °C) as well as the red-hot temperatures of furnaces.

Austenitic stainless steels are commonly recognized as non-magnetic steel and are used for cryogenic applications as well as in the high temperatures of furnaces. This steel is anti-corrosive because it has 16% to 25% chromium, contains nitrogen in solution, nickel and molybdenum. Since this type of stainless steel is anti-corrosive, it can withstand normal corrosive attacks from harsh environmental conditions. It possesses excellent resistance to hot sulfuric acid and many other aggressive environments that would readily attack type 316 stainless steels. This type of steel also provides excellent resistance against stress corrosion cracking when brought in contact with 20-40% boiling sulfuric acid. Also, this stainless steel exhibits excellent mechanical properties and the presence of niobium helps to minimize carbide precipitation during welding.

Chemical composition of austenitic stainless steels Austenitic stainless steels are divided into 5 main groups whose chemical compositions are as follows:

Steel Group	Chemical composition in % (maximum values, unless otherwise indicated)									Notes
	C	Si	Mn	P	S	Cr	Mo	Ni	Cu	
A1	0,12	1,0	6,5	0,200	0.15-0.35	16-19	0,7	5-10	1,75-2,25	2) 3)4)
A2	0,10	1,0	2,0	0,050	0.03	15-20	-	8-19	4	5)6)
A3	0,08	1,0	2,0	0,045	0.03	17-19	-	9-12	1	1)7)
A4	0,08	1,0	2,0	0,045	0.03	16-18.5	2-3	10-15	4	6)8)
A5	0,08	1,0	2,0	0,045	0.03	16-18.5	2-3	10.5-14	4	1)7)8)

1) Stabilized against intergranular corrosion through addition of titanium, possibly niobium, tantalum.

2) Sulfur may be replaced by selenum.

3) If the nickel content is below 8 %, the min. manganese content shall be 5 %.

4) There is no min. limit to the copper content, provided that the nickel content is greater than 8 %.

5) If the chromium content is below 17 %, the min. nickel content should be 12 %.

6) For austenitic stainless steels having a max. carbon content of 0,03 %, nitrogen may be present to a max. of 0,22 %.

7) This shall contain titanium $\geq$ 5 x C up to 0,8 % max. for stabilization and be marked appropriately as specified in this table, or shall contain niobium (columbium) and / or tantalum $\geq$ 10 x C up to 1 % maximum for stabilization and be marked approprately as specified in this table.

8) At the discretion of the manufacturer, the carbon content may be higher where required in order to obtain the specified mechanical properties at larger diameters, but shall not exceed 0,12 % for austenitic steels.

Martensitic Stainless Steel

Martensitic stainless steel is a type of steel having a magnetic, corrosion resistant and hardenable crystalline structure after heat treating. It is composed of chromium deposits with no nickel fractions.

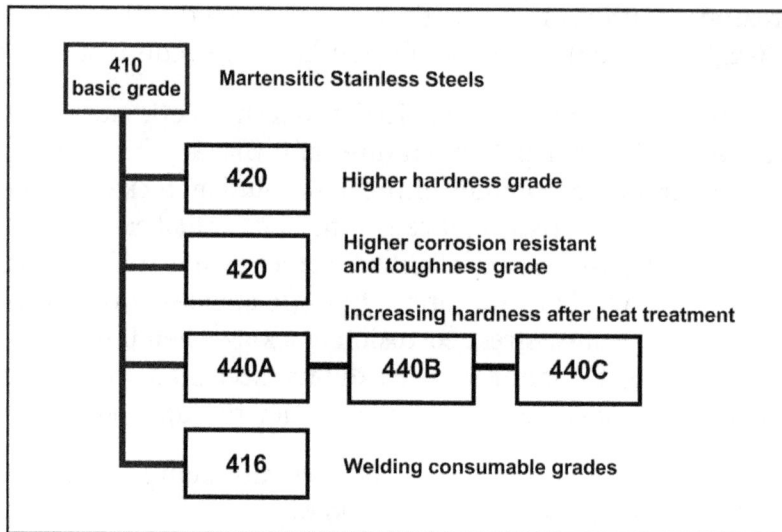

Major grades of martensitic stainless steels

Martensitic grades of stainless steel were developed in order to provide a group of stainless steels which are corrosion resistant and hardenable by heat treatment. Martensitic stainless steels are essentially Fe-Cr-C alloys and are similar to carbon or low alloy steels with a structure similar to the ferritic steels. However, due the addition of carbon, they can be hardened and strengthened by heat treatment, in a similar way to carbon steels. The main alloying elements are chromium (10.5 % to 18 %), molybdenum (0.2 % to 1 %), no nickel (except for two grades), and carbon (0.1 % to 1.2 %). Major grades in the family of martensitic group of stainless steels are given in figure.

Properties

The structures of martensitic stainless steels are body centered tetragonal (bct) and they are classified as a hard ferro magnetic group. In the annealed condition, these steels have tensile yield strengths of around 275 N/sq mm and hence they can be machined, cold formed, or cold worked in this condition. These stainless steels have good ductility and toughness properties, which decrease as strength increases. Martensitic stainless steels can be moderately hardened by cold working. These stainless steels are typically heat treated by both hardening and tempering to yield strength levels up to 1900 N/sq mm. The strength obtained by heat treatment depends on the carbon content of the steels. Increasing the carbon content increases the strength and hardness potential but decreases ductility and toughness. The higher carbon grades are capable of being heat treated to a hardness of 60 HRC.

Martensitic stainless steels may be heat treated, in a similar manner to conventional steels, to provide a range of mechanical properties, but offer higher hardenability and have different heat treatment temperatures. They are subject to an impact transition at low temperatures and possess poor formability. Their thermal expansion and other thermal properties are similar to conventional steels. They may be welded with caution when matching filler metals are used but cracking can be a feature.

All martensitic stainless steels are ferro magnetic. Due to the stresses induced by the hardening transformation, these stainless steels exhibit permanent magnetic properties if magnetized in the hardened condition. For a given grade, the coercive force tends to increase with increasing hardness, rendering these stainless steels more difficult to demagnetize. These stainless steels are not used as permanent magnets to any significant extent.

Cold working increases the coercive force of these steels changing their behaviour from that of a soft magnet to a weak permanent magnet. If parts of cold worked martensitic stainless steel are exposed to a strong magnetic field, the parts can be permanently magnetized and, therefore, able to attract other ferro magnetic objects. Apart from possibly causing handling problems, the parts would be able to attract bits of iron or steel which will, if not removed, impair corrosion resistance. It is therefore prudent to either electrically or thermally demagnetize such parts if they have been subjected to a strong magnetic field during fabrication.

Martensitic stainless steels can be tested by nondestructive testing using the magnetic particle inspection method, unlike austenitic stainless steels.

Optimum corrosion resistance is attained in the heat-treated i.e. hardened and tempered condition. Martensitic stainless steels are less resistant to corrosion. Their corrosion resistance may be described as moderate (i.e. their corrosion performance is poorer in comparison with the austenitic and ferritic grades of stainless steels of the same chromium and alloy content).

The effect of nitrogen on localized corrosion resistance of martensitic stainless steels showed that intergranular corrosion effectively takes place in martensitic microstructures exposed to sulphuric acid solutions, and that nitrogen additions up to 0.2 % [weight %] allow improving resistance to this kind of localized attack.

Martensitic grades of stainless steels can be developed with nitrogen and nickel additions but with lower carbon levels than the traditional grades. These steels have improved toughness, weldability and corrosion resistance.

Heat Treatment

Martensitic stainless steels are usually used in the hardened and tempered condition. The hardening treatment consists of heating to a high temperature in order to produce an austenitic structure with carbon in solid solution followed by quenching. The austenitizing temperature is generally in the range 925 °C to 1070 °C. The effect of austenitizing temperature and time on hardness and strength varies with the composition of the steel, especially the carbon content.

In general the hardness will increase with austenitizing temperature up to a maximum and then decrease. The effect of increased time at the austenitizing temperature normally indicates that there is a slow reduction in hardness with increased time.

Quenching, after austenitizing, is done in air, oil or water depending on steel grade. On cooling below the Ms temperature (starting temperature for the martensite transformation) the austenite transforms to martensite. The Ms temperature lies in the range 300 °C to 700 °C and the transformation is finished at around 150 °C to 200 °C below the Ms temperature.

Almost all alloying elements lower the Ms temperature with carbon having the greatest effect. This means that in the higher alloyed martensitic grades the microstructure contains retained austenite due to the low temperature (below ambient) needed to finish the transformation of the austenite to martensite.

In the hardened condition the strength and hardness are high but the ductility and toughness are low. In order to obtain useful engineering properties, martensitic stainless steels are normally tempered. The tempering temperature used has a large influence on the final properties of the steel.

Usually increasing tempering temperatures below about 400 °C leads to a small decrease in tensile strength and an increase in reduction of area, while hardness, elongation and yield strength are more or less unaffected. Above this temperature there is more or less pronounced increase in yield strength, tensile strength and hardness due to the secondary hardening peak, around 450 °C to 500 °C.

In the temperature range around the secondary hardening peak there is generally a dip in the impact toughness curve. Above about 500 °C there is a rapid reduction in strength and hardness, and a corresponding increase in ductility and toughness. Tempering at temperatures above the 780 °C for the steel, results into partial austenitizing with the possibility of presence of untempered martensite after cooling to room temperature.

Developments in Martensitic Stainless Steels

Martensitic stainless steels are also produced with low carbon content (0.06 % max) and with 3 % to 6 % nickel. These steels, called 'martensitic-austenitic' or 'nickel-martensitic', have a balanced composition that promotes stable austenite after hardening and tempering. They have relatively good weldability.

The low carbon types have been further developed into 'super martensitic' stainless steels. These stainless steels typically contain 11 % to 13.% of chromium, 2 % to 6.% of nickel, 0 % to 3.% of molybdenum and a maximum of 0.030.% of carbon and nitrogen. Their high strength is combined with good impact strength and weldability.

In addition to the standard grades, a large number of alloyed martensitic stainless steels have been developed for moderately high temperature applications. Most common additions include Mo, V and Nb. These lead to a complex precipitation sequence. A small amount (up to 2 weight %) of Ni is added which improves the toughness. The 12Cr-Mo-V-Nb steels are used in the power generation industry, for steam turbine blades operating at temperatures around 600 °C.

Some corrosion-erosion experiments performed with martensitic stainless steels have shown that corrosion-erosion resistance of the high-nitrogen stainless steels is higher than that of the conventional stainless steel for the testing temperatures, in the range from 0 °C to 70 °C. This can be associated to the beneficial effect of nitrogen in solid solution in martensite.

Applications of Martensitic Stainless Steels

Martensitic stainless steels are specified when the application requires good tensile strength, creep, and fatigue strength properties, in combination with moderate corrosion resistance and heat resistance upto approximately 650 °C. Due to their high strength in combination with some corrosion resistance, martensitic steels are suitable for applications where the material is subjected to both corrosion and wear. Martensitic steels with high carbon content are often used for tool steels.

Martensitic stainless steels are used for surgical and dental instruments, wire, screws, springs, razor strips, blades and cutting tools, fasteners, gears and ball bearings and races, gauge blocks, moulds and dies etc. They are also used in the petrochemical industry for steam and gas turbines blades and buckets. Typical other applications are aerospace, automotive, hydroelectric engines, cutlery, defense, power hand tools, pump parts, valve seats, chisels, bushings, shafts, and sporting equipment industry etc.

Many of these applications are hidden to most of the public which probably explains why martensitic stainless steels do not have a prominent public profile. It is good to remind the general public that much of the modern world rests on martensitic stainless steels which is doing their job behind the scenes.

Duplex Stainless Steel

Duplex stainless steel is a type of stainless steel that is composed of grains of two types of stainless steel material, austenitic and ferritic. The word "duplex" refers to the two-phase microstructure of ferritic and austenitic steel grains. The ferritic and austenitic stainless steel grades have approximately equal proportions (i.e., 50% each) in duplex stainless steel.

Duplex stainless steel offers high mechanical strength and excellent corrosion resistance properties.

Duplex stainless steel is considered an alternative to the expensive nickel alloys and high alloy austenitic stainless steel materials used in the most demanding applications.

Metallurgy of Duplex Stainless Steels

During solidification of the DSS melt, the first solid that forms is δ-ferrite. As the temperature drops, austenite formation takes place. After complete solidification, the microstructure is that of austenite islands in a matrix of ferrite. The volume fraction of ferrite-austenite depends on the chemical composition. The large amounts of alloying elements added to DSS results in the formation of various carbides, intermetallics and other secondary phases which form over different temperature ranges at varying rates.

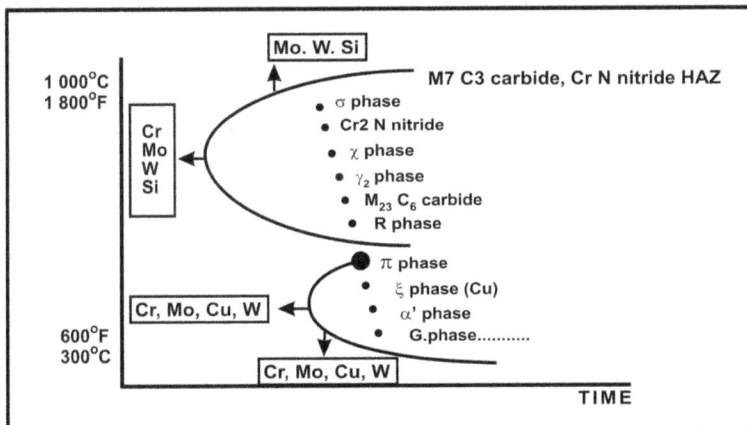

A schematic TTT curve for formation of precipitates in DSS and the effects of alloying elements in the temperature ranges of formation for various precipitates.

Sigma Phase

The Sigma (ζ) phase is a Cr, Mo rich hard embrittling precipitate which forms between 650 and 1000°C is often associated with reduction in impact toughness and corrosion resistance. Since the mobility and concentration of Mo and Cr in ferrite is higher than in austenite, ζ-phase precipitation generally occurs in the ferrite phase. It also forms in the HAZ during welding. It has a tetragonal crystal structure with 32 atoms per unit cell and 5 different crystallographic atom sites. The morphology of ζ-phase changes with temperature. At around 750°C, it has a coral-like structure; at 950°C it is bigger and more compact.

The depletion in Mo content is a lot more pronounced compared to that of Cr, indicating that Mo is the main element controlling the precipitation of ζ-phase. The formation of ζ-phase is rapid and a very high cooling rate is required to avoid it formation during quenching from solutionizing temperature. For a 2205 DSS, a cooling rate of 0.23 K/s is necessary to avoid more than 1% ζ-phase formation.

Chi Phase

The enrichment of ferrite with intermetallic forming elements during long term thermal exposure at temperatures around 700°C, favors the precipitation of Chi (χ) phase. It often nucleates at the δ/γ interface and grows into the δ matrix. It is difficult to study its influence on corrosion and toughness since it often co-exists with ζ-phase. Increase in aging time causes an enrichment of Mo and depletion of Fe in the χ-phase. During isothermal aging, the χ-phase always precipitates before ζ-phase, but during continuous cooling, χ-phase appears only at low cooling rates.

Secondary Austenite

The mechanism and rate of formation of secondary austenite (γ_2) may vary depending on the temperature. In the temperature range of 700–900°C, typical mechanism is by the eutectoid reaction, which is facilitated by rapid diffusion along the δ/γ boundaries giving rise to ζ-phase and γ_2 in prior ferrite grains. This also reduces the Cr and Mo content in the ferrite. When Cr2N precipitates cooperatively, γ_2 has been found to be poor in Cr, making it highly susceptible to pitting corrosion. At temperatures above 650°C, at which diffusion rates are higher, γ_2 is formed as Widmanstätten precipitates. Below 650°C, ferrite transforms to γ_2 by a mechanism quite similar to that of martensite formation. The γ_2 formed in this manner has a similar composition to the ferritic lattice thus indicating that the transformation was diffusionless.

R-phase

Isothermal treatment of duplex stainless steels between 550 and 650°C results in the uniform and very fine distribution of R-phase throughout the δ grains. The R-phase is a Mo rich intermetallic having a trigonal crystal structure. Its formation reduces the toughness and critical pitting temperature in DSS. R-phase precipitates may be intergranular or intragranular in nature; the former perhaps more deleterious with regard to pitting corrosion since they may contain up to 40% Mo. With the increase in aging time, R-phase transforms into ζ-phase due to diffusion of Mo from the R-phase into the ζ-phase, which eventually results in the decrease in the volume fraction of R-phase.

Chromium Nitrides

The solubility of nitrogen at about 1000°C in ferrite is high, but drops on cooling and the ferrite becomes supersaturated in nitrogen, leading to the intergranular precipitation of needle-like Cr_2N.

Isothermal heat treatment in the temperature range of 700–900°C usually results in precipitation of Cr_2N either on the δ/δ grain boundaries or the δ/γ phase boundaries. The hexagonal Cr_2N formed under these conditions has a negative influence on pitting corrosion resistance. In HAZ of welds, however, the cubic Cr_2N is the predominant nitride that has been observed. Cr_2N precipitates display film-like or tiny platelet-like morphology.

Carbides ($M_{23}C_6$ and M_7C_3)

M_7C_3 forms at the δ/γ grain boundaries in the temperature range of 950–1050°C but can be avoided by ordinary quenching methods since its formation takes at least 10 min. $M_{23}C_6$ precipitates rapidly between 650 and 950°C, predominantly at the δ/γ boundaries where Cr-rich ferrite intersects with C-rich austenite. Several precipitate morphologies have been recorded including cuboidal, acicular and cellular form; each having an associated Cr depleted zone in its vicinity. Since modern duplex grades contain less than 0.02%C, carbides of either form are rarely seen.

Alpha Prime

Alpha Prime (α') is a Cr-rich precipitate that forms in the temperature range of 280–525°C. The main cause for formation of α' is the miscibility gap in the Fe-Cr system whereby ferrite undergoes spinodal decomposition into Fe-rich δ-ferrite and Cr-rich α'. Within the miscibility gap but just outside the spinodal, classical nucleation and growth of α' occurs. The α' precipitate has a body-centered crystal structure and is the main cause of hardening and 475°C embrittlement in ferritic stainless steels.

Epsilon Phase

In duplex alloys containing copper, the supersaturation of ferrite due to decrease in solubility at lower temperatures leads to the precipitation of extremely fine particles of Cu-rich Epsilon (ε) phase within the ferrite grains after 100 h at 500°C. This significantly extends the low temperature hardening range for duplex stainless steels. Often, ε-phase has been mistaken for $\gamma 2$ due to similar temperature ranges of formation.

G, π and τ Phases

The G-phase develops at α/α' interfaces between 300 and 400°C after several hours of exposure, due to enrichment of Ni and Si at these locations.

The π-nitride is a Cr and Mo rich precipitate with a cubic crystal structure that forms at intergranular sites in DSS welds after isothermal treatment at 600°C for several hours.

The η-phase is a heavily faulted precipitate with needle-like morphology that forms due to heat treatment in the temperature range of 550–650°C. It has an orthorhombic crystal structure.

Effect of Alloying Elements

Chromium

The primary role of chromium in stainless steels is to improve the localized corrosion resistance, by the formation of a passive Cr-rich oxy-hydroxide film. This film extends the passive range and reduces the rate of general corrosion. The beneficial effect of adding very high levels of chromium is, however, negated by the enhanced precipitation of intermetallic phases which often lead to a reduction in ductility, toughness and corrosion resistance. Apart from this chromium also stabilizes ferrite. Although, other alloying elements can influence the effectiveness of the passive film, none of them can create the properties of stainless steel, by themselves. However, it is often more efficient to improve corrosion resistance by addition of other elements, with or without increasing the chromium content to ensure that the mechanical properties, fabricability, weldability or high temperature stability remain largely unaffected.

Nickel

Nickel, when added in sufficient quantities, stabilizes austenite; this greatly enhances mechanical properties and fabrication characteristics. Nickel effectively promotes re-passivation, especially in reducing environments and is particularly useful in resisting corrosion in mineral acids. Increasing nickel content to about 8–10% decreases resistance to SCC, but on further increase, SCC resistance is restored and is achieved in most service environments at about 30% Ni.

In order to maintain 40–60% ferrite, balance austenite in DSS, the ferrite stabilizing agents need to be balanced with the austenite stabilizers. For this reason, the level of nickel added to a DSS will depend primarily on the chromium content. Excessive nickel contents may enhance intermetallic precipitation when the alloy is exposed to the temperature range of 650–950°C, due to enrichment of ferrite in Cr and Mo. High Ni contents also accelerate α' formation. Although nickel does have some direct effect on corrosion properties, it appears that its main role is to control phase balance and element partitioning.

Molybdenum

Molybdenum, in combination with chromium, effectively stabilizes the passive film in the presence of chlorides. Molybdenum is effective in increasing the resistance to the initiation of pitting and crevice corrosion. Its effect on ferrite stability is similar to that of chromium. To prevent ζ-phase formation in the hot working temperature range, i.e. above 1000°C, the upper limit of Mo addition of is limited to about 4%, since Mo is the main element controlling the precipitation of ζ-phase.

Manganese

Although manganese acts as an austenitic stabilizer in austenitic stainless steels, mixed results have been obtained for DSS in which it has little effect on the phase balance. It appears that Mn increases the temperature range and formation rate of ζ-phase. Manganese increases abrasion and wear resistance and tensile properties of stainless steels without loss of ductility. Further, Mn increases the solubility of nitrogen, thus allowing for higher nitrogen contents. However, Mn additions in excess of 3 and 6%, for nitrogen levels of 0.1 and 0.23% respectively, significantly decrease the critical pitting temperature (CPT). Nevertheless, the combination of Mn and N in modern DSS improves the pitting resistance and counteracts the singular problems associated with Mn.

Nitrogen

Nitrogen enhances pitting resistance by retarding the formation of ζ-phase and diminishes Cr and Mo segregation and also raises the corrosion resistance of the austenitic phase in DSS. Nitrogen has also been reported to increase crevice corrosion resistance. Nitrogen strengthens austenite by dissolving at the interstitial sites in solid solution. Addition of nitrogen to DSS, suppresses austenite dissolution and encourages austenite reformation in the HAZ.

Copper

Copper reduces the corrosion rate of high alloy austenitic grades in non-oxidizing environments, such as sulfuric acid. In some DSS with 25% Cr, 1.5% Cu is added to obtain the optimum corrosion resistance in 70% H_2SO_4 at 60°C. For boiling HCl, an addition of 0.5% Cu decreases both active dissolution and crevice corrosion rates. Copper additions in DSS are limited to about 2%, since higher levels, reduce hot ductility and can lead to precipitation hardening. Exposure at temperatures between 300 and 600°C can lead to precipitation of tiny Cu-rich precipitates which do not significantly reduce corrosion resistance or toughness but can be exploited for improving abrasion-corrosion resistance in duplex pump castings.

Tungsten

Up to 2% tungsten additions have been made in DSS to improve pitting resistance. Tungsten also increases crevice corrosion resistance in heated chloride solutions. Tungsten encourages intermetallic formation in the 700–1000°C temperature range, and also encourages γ_2 formation in weld metal. Thermodynamically, it is believed to be equivalent to Mo with respect to ζ-phase formation, but this is not the case in terms of kinetics. Levels between 1 and 3% have been shown to restrict ζ-phase formation to the intergranular sites instead of phase boundaries; the influence of the large tungsten atom on the diffusion of Mo and W at the phase boundaries is thought to be the reason. Tungsten alloyed weld metal has been shown to form χ-phase more rapidly than in welds without W additions. Generally, the tungsten content is limited to 1% in a 4% Mo DSS, and 2% in DSS with about 3%.

Silicon

Silicon enhances high temperature oxidation resistance and is also beneficial for concentrated nitric acid service. DSS bearing high silicon (3.5–5.5%) have enhanced pitting corrosion resistance and a claimed immunity to SCC. However, it is preferred to limit Si additions to 1% since Si is generally considered to enhance ζ-phase formation.

Carbon, Sulfur and Phosphorous

The carbon content of most wrought DSS is limited to 0.02 to 0.03%, primarily to suppress precipitation of Cr-rich carbides along the grain boundaries. Sulfur and phosphorous contents are controlled but not eliminated. The presence of S is important for weld bead penetration. Modern steel making processes such as argon oxygen decarburization (AOD) and vacuum oxygen decarburization (VOD) help in controlling the levels of S and C, while P contents can be reduced by using good melting practice.

Effect of Heat Treatment

Solution Annealing

Element solubility in ferrite falls with decreasing temperature, increasing the probability of precipitation during heat treatment. During solidification, DSS solidifies completely as ferrite and then undergo solid state transformation into austenite. This is a reversible process and as a result, any large increases in temperature above 1000°C lead to an increase in ferrite and also a reduction in the partitioning of substitutional elements between phases. In addition, ferrite becomes enriched in interstitial elements such as carbon and nitrogen.

Heat treatment in the temperature range 1100–1200°C can have a dramatic influence on the microstructure of a wrought product. Prolonged treatment at high temperatures can lead to equiaxed grains, whereas, cooling at intermediate rates can render the grains acicular, with Widmanstätten morphology. Step quenching, with or without simultaneous mechanical strain can lead to a dual structure, consisting of both coarse and fine austenite grains.

Duplex alloys with high contents of Cr, Mo and W are most susceptible to intermetallic precipitation. Mo and W extend the stability range of intermetallics to higher temperatures. For this reason, higher solution annealing temperatures, i.e. above 1000°C are necessary. In order for the precipitates to re-dissolve, solution annealing temperatures for superduplex grades must be performed at 1050°C and above. For grades such as S32550 and S32750, a few minutes at 1050–1070°C are sufficient, whereas, for tungsten bearing grades such as S32760, 1100°C has been recommended. Li et al., during their investigation of a hyper duplex stainless steel without W additions, solution annealed the samples at 1100°C for 1 h and obtained a microstructure of ferrite and austenite with no secondary phases. Jeon et al. 2012 solution treated a hyper duplex alloy with high W additions at 1090°C for 30 min and found no secondary phases.

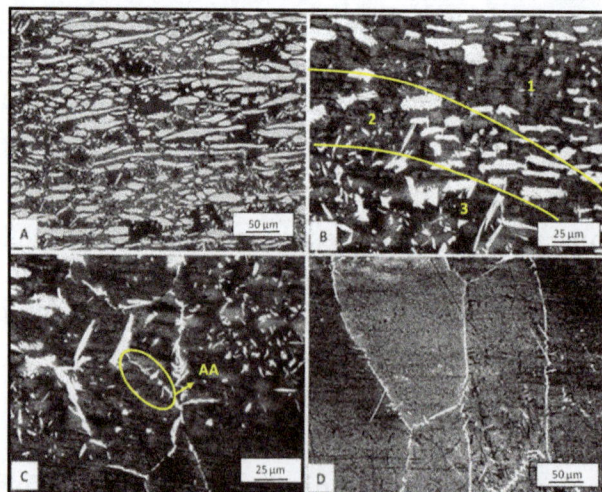

Microstructure of as received S2205 duplex stainless steel sample (Etchant:Beraha's tint etch).

Figure above shows the SEM micrograph of the as-received S2205 DSS in the wrought form in which austenitic islands are embedded in the ferritic matrix with some undissolved precipitates. Hence, in order to dissolve these harmful precipitates samples were subjected to solution heat treatment by heating it to 1110°C for 60 min. The solution heat treatment was also done to adjust the austenite and ferrite phase proportions which can be seen in the figure. If any macro segregations are present in the sample the solution heat treatment will help to eliminate them.

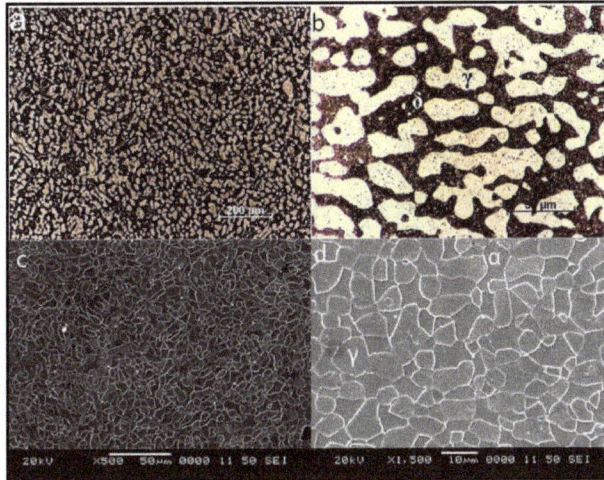

Microstructure of S2205 after solution heat treatment (a) OM image of distribution of austenite grains in the ferrite matrix (magnification: 100×). (b) OM image of phase morphology (magnification: 500×).

Advantages of Duplex Stainless Steel

Based on the above behavior, dual phase stainless steel has many advantages:

Excellent Resistance to Neutral Chloride Stress Corrosion

The duplex stainless steel with chromium content in the range of 18% -22% is under low stress and has excellent resistance to neutral chloride stress corrosion. 18 - 8 type austenitic stainless steel is usually used in more than 70 °C in neutral chloride solutions to stress corrosion cracking, heat exchanger, evaporator and other equipment manufactured by the stainless steel in trace chloride and hydrogen sulfide in industrial environment are susceptible to stress corrosion cracking, but duplex stainless steel in this resistance performance is good.

Good Corrosion Resistance

Duplex stainless steel, if the alloy element molybdenum is added, will form better pitting corrosion resistance. When pitting equivalent value is equal, the critical pitting corrosion potential of duplex stainless steel and austenitic stainless steel is close. The corrosion resistance of duplex stainless steel with Cr content of 18% is equivalent to that of AISI316L stainless steel. The pitting corrosion and crevice corrosion resistance of duplex stainless steels containing 25%Cr, especially those containing oxygen, are even higher than those of AISI316L stainless steel.

Good Corrosion Resistance, Fatigue and Corrosion Resistance

Duplex stainless steel has good corrosion resistance, fatigue and wear and corrosion resistance, certain conditions have also been used for pumps, valves and other equipment to manufacture raw materials.

The Overall Mechanical Properties are Good

The overall mechanical properties of duplex stainless steel are better, the strength and fatigue strength are higher, and the yield strength is two times of type 18-8 austenitic stainless steel.

Good Weldability

The weldability of duplex stainless steel is good and the tendency of hot cracking is small. There is usually no need to preheat before welding, without heat treatment after welding, and can be welded to dissimilar steels such as type 18-8 austenitic stainless steel or carbon steel.

The Range of Hot Working Temperature is High

The hot working temperature range of duplex stainless steel containing low chromium (18%Cr) is wider than that of type 18-8 austenitic stainless steel, and the resistance is small. The steel sheet can be rolled directly without forging. The hot work of duplex stainless steel containing high chromium (25%Cr) is more difficult than austenitic stainless steel.

Precipitation Hardening Stainless Steels

Precipitation hardening stainless steels are chromium and nickel containing steels that provide an optimum combination of the properties of martensitic and austenitic grades. Like martensitic grades, they are known for their ability to gain high strength through heat treatment and they also have the corrosion resistance of austenitic stainless steel.

The high tensile strengths of precipitation hardening stainless steels come after a heat treatment process that leads to precipitation hardening of a martensitic or austenitic matrix. Hardening is achieved through the addition of one or more of the elements Copper, Aluminium, Titanium, Niobium, and Molybdenum.

The most well known precipitation hardening steel is 17-4 PH. The name comes from the additions 17% Chromium and 4% Nickel. It also contains 4% Copper and 0.3% Niobium. 17-4 PH is also known as stainless steel grade 630.

The advantage of precipitation hardening steels is that they can be supplied in a "solution treated" condition, which is readily machinable. After machining or another fabrication method, a single, low temperature heat treatment can be applied to increase the strength of the steel. This is known as ageing or age-hardening. As it is carried out at low temperature, the component undergoes no distortion.

Characterisation

Precipitation hardening steels are characterised into one of three groups based on their final microstructures after heat treatment. The three types are: martensitic (e.g. 17-4 PH), semi-austenitic (e.g. 17-7 PH) and austenitic (e.g. A-286).

Martensitic Alloys

Martensitic precipitation hardening stainless steels have a predominantly austenitic structure at annealing temperatures of around 1040 to 1065°C. Upon cooling to room temperature, they undergo a transformation that changes the austenite to martensite.

Semi-austenitic Alloys

Unlike martensitic precipitation hardening steels, annealed semi-austenitic precipitation hardening steels are soft enough to be cold worked. Semi-austenitc steels retain their austenitic structure at room temperature but will form martensite at very low temperatures.

Austenitic Alloys

Austenitic precipitation hardening steels retain their austenitic structure after annealing and hardening by ageing. At the annealing temperature of 1095 to 1120°C the precipitation hardening phase is soluble. It remains in solution during rapid cooling. When reheated to 650 to 760°C, precipitation occurs. This increases the hardness and strength of the material. Hardness remains lower than that for martensitic or semi-austenitic precipitation hardening steels. Austenitic alloys remain nonmagnetic.

Strength

Yield strengths for precipitation-hardening stainless steels are 515 to 1415 MPa. Tensile strengths range from 860 to 1520 MPa. Elongations are 1 to 25%. Cold working before ageing can be used to facilitate even higher strengths.

Chemical Composition

1.4542 Steel	Spec: EN 10088-3:2005
Chemical Element	% Present
Carbon (C)	0.0 - 0.07
Chromium (Cr)	15.00 - 17.00
Manganese (Mn)	0.0 - 1.50
Silicon (Si)	0.0 - 0.70
Phosphorous (P)	0.0 - 0.04
Sulphur (S)	0.0 - 0.03
Nickel (Ni)	3.00 - 5.00
Copper (Cu)	3.00 - 5.00
Molybdenum (Mo)	0.0 - 0.60
Niobium (Columbium) (Nb)	0.0 - 0.45
Iron (Fe)	Balance

Strength of Precipitation-hardening Stainless Steels

Yield strengths for precipitation-hardening stainless steels are 515 to 1415 MPa. Tensile strengths range from 860 to 1520 MPa. Elongations are 1 to 25%. Cold working before ageing can be used to facilitate even higher strengths.

Heat Treatment of Precipitation-hardening Stainless Steels

The key to the properties of precipitation hardening stainless steels lies in heat treatment.

After solution treatment or annealing of precipitation hardening stainless steels, a single low temperature "age hardening" stage is employed to achieve the required properties. As this treatment is carried out at a low temperature, no distortion occurs and there is only superficial discolouration. During the hardening process a slight decrease in size takes place. This shrinking is approximately 0.05% for condition H900 and 0.10% for H1150.

Typical mechanical properties achieved for 17-4 PH after solution treating and age hardening are given in the following table. Condition designations are given by the age hardening temperature in °F.

Table: Mechanical property ranges after solution treating and age hardening.

Cond.	Hardening Temp and time	Hardness (Rockwell C)	Tensile Strength (MPa)
A	Annealed	36	1100
H900	482°C, 1 hour	44	1310
H925	496°C, 4 hours	42	1170-1320
H1025	552°C, 4 hours	38	1070-1220
H1075	580°C, 4 hours	36	1000-1150
H1100	593°C, 4 hours	35	970-1120
H1150	621°C, 4 hours	33	930-1080

Typical Chemical Composition of Precipitation-hardening Stainless Steels

Table: Typical chemical composition for stainless steels alloy 17-4PH.

	17-4 PH
C	0.0-0.07
Mn	0.0-1.5
Si	0.0-0.7
P	0.0-0.04
S	0.0-0.02
Cr	15.0-17.0
Ni	3.0-5.0
Cu	3.0-5.0
Nb	0.0-0.45
Mo	0.0-0.6
Fe	Balance

Typical Mechanical Properties of Precipitation-hardening Stainless Steels

Table: Typical mechanical properties for stainless steels alloy 17-4PH.

Grade 17-4PH	Annealed	Cond 900	Cond 1150
Tensile Strength (MPa)	1100	1310	930
Elongation A5 (%)	15	10	16
Proof Stress 0.2% (MPa)	1000	1170	724
Elongation A5 (%)	15	10	16

Typical Physical Properties of Precipitation-hardening Stainless Steels

Table: Typical physical properties for stainless steels alloy 17-4PH.

Property	Value
Density	7.75 kg/m^3
Modulus of Elasticity	196 GPa
Electrical Resistivity	$0.080 \times 10^{-6} \ \Omega.\text{m}$
Thermal Conductivity	18.4 W/m.K
Thermal Expansion	$10.8 \times 10^{-6} \ /\text{K}$

Alloy Designations

Stainless steels 17-4 PH also corresponds to a number of following standard designations and specifications.

Table: Alternate designations for stainless steels alloy 17-4PH

Euronorm	UNS	BS	En	Grade
1.4542	S17400	-	-	630

Corrosion Resistance of Precipitation-hardening Stainless Steels

Precipitation hardening stainless steels have moderate to good corrosion resistance in a range of environments. They have a better combination of strength and corrosion resistance than when compared with the heat treatable 400 series martensitic alloys. Corrosion resistance is similar to that found in grade 304 stainless steels.

In warm chloride environments, 17-4 PH is susceptible to pitting and crevice corrosion. When aged at 550°C or higher, 17-4 PH is highly resistant to stress corrosion cracking. Better stress corrosion cracking resistance comes with higher ageing temperatures.

Corrosion resistance is low in the solution treated (annealed) condition and it should not be used before heat treatment.

Heat Resistance of Precipitation-hardening Stainless Steels

17-4 PH has good oxidation resistance. In order to avoid reduction in mechanical properties, it should not be used over its precipitation hardening temperature. Prolonged exposure to 370-480°C should be avoided if ambient temperature toughness is critical.

Fabrication of Precipitation-hardening Stainless Steels

Fabrication of all stainless steels should be done only with tools dedicated to stainless steel materials or tooling and work surfaces must be thoroughly cleaned before use. These precautions are necessary to avoid cross contamination of stainless steels by easily corroded metals that may discolour the surface of the fabricated product.

Cold Working of Precipitation-hardening Stainless Steels

Cold forming such as rolling, bending and hydroforming can be performed on 17-4PH but only in the fully annealed condition. After cold working, stress corrosion resistance is improved by re-ageing at the precipitation hardening temperature.

Hot Working of Precipitation-hardening Stainless Steels

Hot working of 17-4 PH should be performed at 950°-1200°C. After hot working, full heat treatment is required. This involves annealing and cooling to room temperature or lower. Then the component needs to be precipitation hardened to achieve the required mechanical properties.

Machinability

In the annealed condition, 17-4 PH has good machinability, similar to that of 304 stainless steels. After hardening heat treatment, machining is difficult but possible.

Carbide or high speed steel tools are normally used with standard lubrication. When strict tolerance limits are required, the dimensional changes due to heat treatment must be taken into account.

Welding of Precipitation-hardening Stainless Steels

Precipitation hardening steels can be readily welded using procedures similar to those used for the 300 series of stainless steels.

Grade 17-4 PH can be successfully welded without preheating. Heat treating after welding can be used to give the weld metal the same properties as for the parent metal. The recommended grade of filler rods for welding 17-4 PH is 17-7 PH.

Applications of Precipitation-hardening Stainless Steels

Due to the high strength of precipitation hardening stainless steels, most applications are in aerospace and other high-technology industries.

Applications include:

- Gears

- Valves and other engine components
- High strength shafts
- Turbine blades
- Moulding dies
- Nuclear waste casks

Structural Steel

Structural steel is a type of steel that is used for construction purposes, and this type of steel is available in a variety of shapes. Structural steel can vary by shape, size, chemical composition, and mechanical properties.

Structural Steel Shapes

I-Beam

An I-Beam (or H-Beam) is a structural steel with an I or H shaped cross-section. The horizontal part of the beam is called a flange, and the vertical part is called a web. The web resists the shear forces, and the flange resists most of the bending moment that the beam experiences. This type of structural steel is a popular choice in construction due to its efficiency in resisting bending and shear loads.

Structural Channel

A structural channel (or C-Beam) is similar to an I-Beam as it has both flanges and webs, but the flanges only stick out of one side of the web, creating a C shaped cross-section. They are often used in building construction and civil engineering where the back side of the web can be mounted against another flat surface for a maximum contact area.

HSS

A hollow structural section (HSS) is a type of structural steel that has a hollow tubular cross section. The term HSS is commonly used in the USA, but in countries that follow British construction terminology the terms circular hollow section (CHS), square hollow section (SHS) or rectangular hollow section (RHS). These three names reference the three basic shapes that HSS steel can be supplied in.

Rectangular sections of HSS steel are commonly utilized in welded steel frames where loading is experienced in multiple directions. Circular and square HSS steel is often used for multiple-axis loading as their uniform geometry along multiple cross-sectional axis provided uniform strength characteristics.

Properties of Structural Steel

Properties of structural steel include:

- Tensile properties
- Shear properties
- Hardness

- Creep
- Relaxation
- Fatigue

Tensile Properties of Structural Steel

There are different categories of steel structures which can be used in the construction of steel buildings. Typical stress strain curves for various classes of structural steel, which are derived from steel tensile test, are shown in figure.

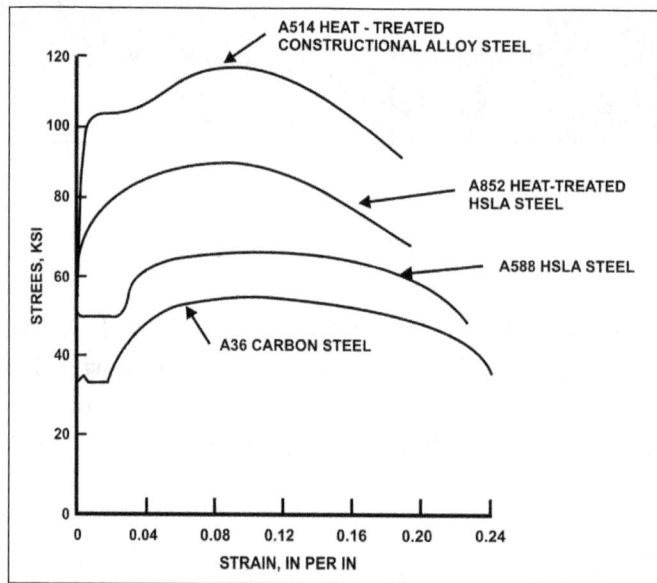

Typical Stress Strain Curve Different Classes of Structural Steel.

The initial part of the curve represents steel elastic limit. In this range, steel structure deformation is not permanent, and the steel regain its original shape upon the removal of the load.

The elastic modulus of all steel classes is same and equal to 200000MPa or 2×10^6 MPa. As the load on the steel is increased, it would yield at a certain point after which plastic range will be reached.

The yield point is the point at which steel specimen reach 0.002 strain under the effect of specific stress (yield stress).

Ductility of steel structure as shown in figure is crucial properties that allow redistribution of stress in continuous steel elements. Ductility is expressed by percentage of steel cross sectional reduction.

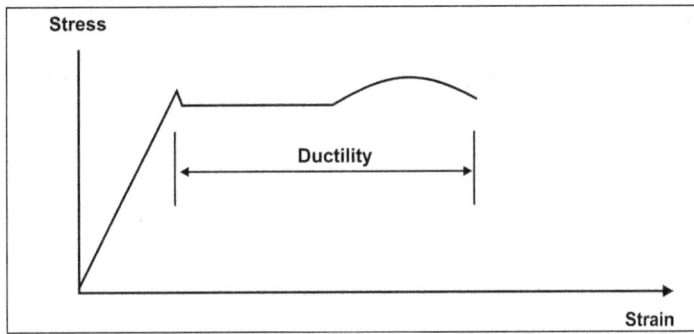

Stress Strain Curve of Structural Steel.

As far as poisons ratio is concerned, it is the ratio of transverse strain to axial strain and it is about 0.30 and 0.50 in elastic and plastic range, respectively.

Regarding cold working of structural steel, it is the process in which different shapes of steel structure are produced at room temperature.

Consequently, steel structure ductility is increased but its ductility is reduced. Residual stress is a stress that stays in steel element after it has been fabricated.

It is necessary to consider strain rate while tensile test is conducted because it modifies steel tensile properties.

If steel structure is used for dynamic loads, then high strain rate would be considered. However, normal strain rate is adopted for steel used in the construction of structure designed for static loads.

The ability of steel structure to accommodate energy is called steel toughness.

Shear Properties of Structural Steel

Shear strength of steel structure is specified at the failure under shear stress and it is about 0.57 times yield stress of structural steel.

Regarding elastic shear modulus, it is expressed as the ratio of shear stress to shear strain in elastic range of steel structure.

Commonly, elastic shear modulus of steel structure can be taken as 75.84Gpa or the following formula can be used to compute elastic shear modulus.

$$G = \frac{E}{2(1-\mu)} \rightarrow \text{Equation-1}$$

Where:

- G: Steel structure shear elastic modulus
- E: Modulus of elasticity of steel structure
- Position's ratio of structural steel: Position's ratio

Hardness of Structural Steel

Hardness is the measure of ability of steel structure to withstand inelastic deformation. Standard test methods and definitions for mechanical testing of steel products (A370-05) specify three different tests to evaluate steel hardness namely: Brinell, Rockwell and portable.

Any of these tests can be used to estimate steel structure hardness. Not only is the steel structure hardness used to examine the uniformity of different products but also to evaluate steel tensile strength.

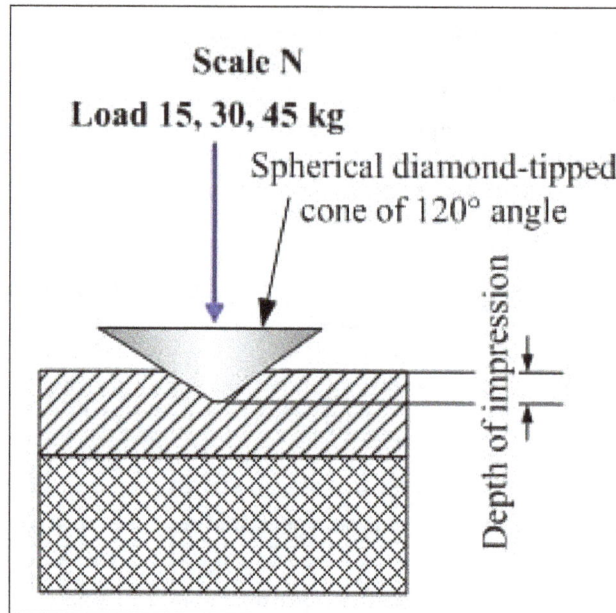

Rockwell Test for Structural Steel Hardness Evaluation.

Creep of Structural Steel Relaxation

Creep is gradual variation of strain of steel structure under constant stress. It occurs due to the influence of constant stress and the effect of fire.

Creep property is insignificant for structural steel frame design and construction apart from the case in which the effect of fire should be taken into consideration.

Structural Steel Relaxation

It is a step by step reduction of structural steel under a constant stress. Usually, yield strength of steel structure increases around 5% over stress relieved strain and the steel structure would suffer from plastic elongation which around 0.01.

Fatigue of Structural Steel

Fatigue is the failure of steel structure due to crack initiation and development under the influence of cyclic loading. Various tests are available to evaluate structural steel fatigue such as flexure test, rotating beam test and axial load test.

Fatigue Test of Structural Steel.

Applications of Structural Steel

Construction

The construction industry utilizes a lot of structural steel for obvious reasons. The high tensile and shear strength offered by the different sections of structural steel enable it to be utilized in the construction of buildings, ensuring a secure load.

Grades including S355JR, S355AR, S355J2/K2+N, S275JR, and S275AR are popular choices for the construction industry due to their high yield strength and versatility of use.

Offshore

Offshore structures are exposed to some of the harshest conditions, including strong winds, salt water, and powerful sea currents. Structural steel is a popular choice for the offshore industry as it promotes a safe environment, longer working life, and reduces the risk of failure.

Shipbuilding

Structural steel is a popular choice for the shipbuilding industry as it can provide a high level of strength, which helps to reduce the overall costs. Structural steel can also reduce maintenance due to the corrosion resistance properties.

Structural Steel from Masteel

Masteel supply a range of different structural steels to a variety of global industries. Our materials have high strength and high yield, and we can supply them with various finishes and testing options.

Tool Steel

Tool steel refers to a variety of carbon and alloy steels that are particularly well-suited to be made into tools. Their suitability comes from their distinctive hardness, resistance to abrasion and deformation and their ability to hold a cutting edge at elevated temperatures. As a result tool steels are suited for their use in the shaping of other materials.

There are six groups of tool steel: water-hardening, cold-work tool steels, shock-resisting, high-speed, hot-work, and special purpose/plastic mold tools steel. The choice of group to select depends on, cost, working temperature, required surface hardness, strength, shock resistance, and toughness requirements. The more severe the service condition (higher temperature, abrasiveness, corrosiveness, loading), the higher the alloy content and consequent amount of carbides required for the tool steel.

Water-hardening Group

Named from its essential property of having to be water quenched. This group of tool steel is essentially plain high carbon steel. It is commonly used because of its low cost.

W-group tool steel gets its name from its defining property of having to be water quenched. W-grade steel is essentially high carbon plain-carbon steel. This group of tool steel is the most commonly used tool steel because of its low cost compared to others. They work well for small parts and applications where high temperatures are not encountered; above 150 °C (302 °F) it begins to soften to a noticeable degree. Its hardenability is low, so W-group tool steels must be subjected to a rapid quenching, requiring the use of water. These tool steels can attain high hardness (above HRC 66) and are rather brittle compared to other tool steels. W-steels are still sold, especially for springs, but are much less widely used than they were in the 19th and early 20th centuries. This is partly because W-steels warp and crack much more during quench than oil-quenched or air hardening steels.

Cold-work Group

This is a group of three tool steels: oil-hardening, air-hardening, and high carbon-chromium. The steels in the group have high hardenability and wear resistance, with average toughness. Typically they are in the production of larger parts or parts that have a minimum distortion requirement when being hardened.

Both Oil quenching and Air-hardening both reduce the distortion and higher stress caused by the quick water quenching.Because of this they are less likely to crack.

Oil-hardening

A very common oil hardening steel is O1 steel. It is a very good cold work steel and also makes very good knives and forks. It can be hardened to about 57-61 HRC.

Air-hardening

The first air-hardening grade tool steel was mushet steel, which was known as air-hardening steel at the time.

Modern air-hardening steels are characterized by low distortion during heat treatment because of their high-chromium content. Their machinability is good and they have a balance of wear resistance and toughness (i.e. between the D- and shock-resistant grades).

High Carbon-chromium

D-type

The D-type, of the cold-work class of tool steels, contain between 10% and 13% chromium. These steels retain their hardness up to a temperature of 425 °C (797 °F). Common applications for these tool steels include forging dies, die-casting die blocks, and drawing dies. Due to their high chromium content, certain D-type tool steels are often considered stainless or semi-stainless, however their corrosion resistance is very limited due to the precipitation of the majority of their chromium and carbon constituents as carbides.

D2 Tool Steel is very wear resistant but not as tough as lower alloyed steels. The mechanical properties of D2 are very sensitive to heat treatment. It is widely used for the production of shear blades, planer blades and industrial cutting tools; sometimes used for knife blades.

A	A2 – A10	Air hardening, Medium alloys
D	D2 – D7	High carbon, high chromium
O	O1 – O7	Oil hardening, Low carbon

ISO 1.2767, also known as DIN X 45 NiCrMo 4, AISI 6F7, and BS EN 20 B, is an air-hardening tool steel with a primary alloying element of nickel. It possesses good toughness, stable grains, and is highly polishable. It is primarily used for dies in plastic injection molding application that involve high stresses. Other applications include blanking dies, forging dies, and industrial blades.

Table: The following steel grades are also in cold work steel group.

AISI Code	AISI Designation	Type of Tool steel
W	W1 to W7	High carbon water hardening steel
	W1 A – 1B	Carbon
	W2 – W3	Carbon Vanadium
	W4 – W5	Carbon Chromium
	W7	Carbon Chromium Vanadium

Shock-resisting Group

This class has high shock resistance and good hardenability. It is designed to resist shock at both low and high temperatures. It also has a very high impact toughness and relatively low abrasion resistance.

Carbide-forming alloys provide the necessary abrasion resistance, hardenability, and hot-work characteristics. This family of steels displays very high impact toughness and relatively low abrasion resistance and can attain relatively high hardness (HRC 58/60). In the US, toughness usually derives from 1 to 2% silicon and 0.5-1% molybdenum content. In Europe, shock steels often contain 0.5-0.6 % carbon and around 3% nickel. 1.75% to 2.75% nickel is still used in some shock resisting and high strength low alloy steels (HSLA), such as L6, 4340, and Swedish saw steel, but it is relatively expensive. An example of its use is in the production of jackhammer bits.

High Speed Group

T-type and M-type tool steels are used for cutting tools when strength and hardness must be retained at high temperatures.

High-speed steel (HSS or HS) is a subset of tool steels, commonly used in tool bits and cutting tools. It is often used in power-saw blades and drill bits. It is superior to the older high-carbon steel tools used extensively through the 1940s in that it can withstand higher temperatures without losing its temper (hardness). This property allows HSS to cut faster than high carbon steel, hence the name high-speed steel. At room temperature, in their generally recommended heat treatment, HSS grades generally display high hardness (above HRC60) and abrasion resistance (generally linked to tungsten and vanadium content often used in HSS) compared with common carbon and tool steels.

M	M1, M7, M10	Molybdenum
	M30, M33, M34, M42, M43, M46, M47	Molybdenum, Cobalt
	M2, M3, M4	Molybdenum, Tungsten
	M6, M15, M35, M36< M41, M44, M45	Molybdenum, Tungsten, Cobalt
T	T1, T2, T3, T7, T9	Tungsten
	T4, T5, T6, T8, T15	Tungsten, cobalt

Hot-working Group

H-group tool steels were specifically developed to maintain strength and hardness while exposed to prolonged elevated temperatures.

Hot-working steels are a group of steel used to cut or shape material at high temperatures. H-group tool steels were developed for strength and hardness during prolonged exposure to elevated temperatures. These tool steels are low carbon and moderate to high alloy that provide good hot hardness and toughness and fair wear resistance due to a substantial amount of carbide. H1 to H19 are based on a chromium content of 5%; H20 to H39 are based on a tungsten content of 9-18% and a chromium content of 3–4%; H40 to H59 are molybdenum based.

H	H 10, H11, h12, H13	Chromium, Molybdenum
	H14, H16, H19, H23	Chromium, Tungsten
	H20, H21, H22, H24, H25, H26	Tungsten
	H15, H41, H42, H43	Molybdenum

Special Purpose Group / Plastic Mold Steel

P-type tool steel is short for plastic mold steels. They are designed to meet the requirements of zinc die casting and plastic injection molding dies. Common steel grades like P20, 420 etc.

L-type tool steel is short for low alloy special purpose tool steel. L6 is extremely tough.

F-type tool steel is water hardened and substantially more wear resistant than W-type tool steel.

F	F1	High carbon, low alloys
	F2, F3	Tungsten
L	L1, L3, L7	Carbon > 0.65%, Chromium
	L2	Carbon <0.65%, Chromium
	L6	Carbon > 0.65%, Nickel
S	S1, S3	Tungsten
	S2, S4, S5, S6	Silicon
	S7	Chromium
P	P1- P20, P21	Low carbon mold steel

Tool steels are metallurgically "clean," high-alloy steels that are melted in relatively small heats in electric furnaces and produced with careful attention to homogeneity. They can be further refined by argon/oxygen decarburization (AOD), vacuum methods, or electroslag refining (ESR). As a result, tool steels are often specified for critical high-strength or wear-resistant applications. Because of their high alloy content, tool steels must be rolled or forged with care to produce satisfactory bar products.

To develop their best properties, tool steels are always heat treated. Because the parts may distort during heat treatment, precision parts should be semifinished, heat treated, then finished. Severe distortion is most likely to occur during liquid quenching, so an alloy should be selected that provides the needed mechanical properties with the least severe quench.

Manufacturing Processes for Tool Steel

The manufacture of tool steels takes place under carefully controlled conditions to produce the required quality. Tool steel has a carbon content of between 0.5% and 1.5%. The manufacturing process introduces alloying elements that form carbides, commonly tungsten, chromium, vanadium and molybdenum.

The most important manufacturing processes for tool steel are as follows:

- Primary Melting
- Electroslag Melting
- Primary Breakdown
- Rolling

- Hot and Cold Drawing
- Continuous Casting
- Powder Metallurgy
- Osprey Process

Primary Melting

Tool steel is often made from around 75% scrap – a mixture of mill scrap and purchased scrap. It's very important to avoid contamination of the scrap, especially from metals which cannot be oxidized like nickel, cobalt and copper.

The majority of tool steel production is done through Electric Arc Furnace (EAF) melting. There are two Stages:

- The scrap is melted rapidly in the furnace.

- The hot metal is transferred to a separate ladle or converter vessel to be refined. This process is known as secondary refining, and it allows for great efficiency and the processing of large volumes.

The refined metal is then transferred into the casting station and poured into ingots. The resulting ingots are usually annealed (heated and cooled slowly) to prevent cracking.

Electroslag Melting

Electroslag remelting or refining (ESR) is a progressive melting process used to produce ingots with smooth surfaces and no pipe (holes) or porosity (imperfections). ESR ingots give improved hot workability, better processing yields, increased cleanliness, better transverse tensile ductility and fatigue properties.

ESR is an expensive process, and the costs saved through the increase in yield are not always sufficient to offset the costs of ESR processing. However for some specialized tool steel applications ESR is worth it.

Vacuum arc remelting (VAR) is a process sometimes used alongside ESR. However its use in tool steels is limited to specialized applications with specific bearing requirements. In the VAR process, heat is supplied via an arc in a high-vacuum environment. The resulting steel has a refined macrostructure and microstructure and excellent chemical uniformity.

Primary Breakdown

The breakdown method used for tool steels employs either an open-die hydraulic press or rotary forging machine. These processes are extremely versatile and can produce lengths of 6 to 13 m (20 to 43 ft) in squares, rectangles, hollows or stepped cross sections. The final product is very high quality, having few cracks, laps or seams, and a high degree of straightness can be achieved.

Rolling

In modern steel manufacture, up to 26 rolling mills are used in a row. The metal is heated via a gas-fired pusher, walking-beam furnace, or high powered induction furnace. Rapid heating is used to prevent decarburization (loss of carbon content). The process is automated by computers and measuring devices are used to monitor the diameter tolerance and surface quality of the metal. Through this process, a coil of steel sheet can be produced in less than 12 minutes.

Hot and Cold Drawing

Drawing operations are usually used on tool steels to produce better tolerances, smaller sizes, or special shapes. As tool steels are of high strength and limited ductility, cold drawings are limited to a single light pass in order to prevent breakage. Warm drawing at temperatures up to 540 °C (1000 °F) is used in multiple passes to strengthen the metal.

Continuous Casting

Continuous casting of tool steel is sometimes done for economic reasons. Following casting, the billets are annealed and sometimes ground, then forged by hammer or rotary, after which they can be rolled. Electroslag rapid remelting (ESRR) is a modern process which runs at higher temperatures than ESR.

Powder Metallurgy

Powder metallurgy (P/M) is used to produce highly alloyed steels such as high-carbon, high-chromium and high-speed. This process has become increasingly popular in recent years. Using traditional methods, the production of high-carbon, high-alloy tool steels are particularly challenging. The relatively slow cooling times for these methods results in the formation of undesirable coarse structures of eutectic carbide, which results in non-uniform heat-treat response, poor transverse qualities and low toughness.

In P/M, the problems of traditional methods are overcome. A fine, uniform distribution of carbides can be produced using P/M which results in improved machinability in the annealed condition, a faster response to hardening heat treatment, and improved grindability. However, there is a downside — a reduction in wear resistance.

Osprey Process

The Osprey Process remains a very specialized activity limited to sites in Japan and the UK, However, it has tremendous technical and commercial potential. The molten alloy is poured from an induction furnace through a nozzle and blasted with high-pressure gas atomization jets, causing the formation of small droplets. The droplets are collected and used to form billets, hollows and sheets.

The advantages of the Osprey process are similar to P/M. Tool steel produced from Osprey material have a uniform distribution of fine carbides. However, the Osprey process is currently not as economically competitive as P/M.

References

- Steel, technology: britannica.com, Retrieved 10 June, 2019

- 30476-what-is-heat-treatment, manufacturing-technology: brighthubengineering.com, Retrieved 3 January, 2019

- Carbonsteel: princeton.edu, Retrieved 14 April, 2019

- Carbon-steel, engineering: sciencedirect.com, Retrieved 4 January, 2019

- Knowles, Peter Reginald (1987), Design of structural steelwork (2nd ed.), Taylor & Francis, p. 1, ISBN 978-0-903384-59-9

- The-properties-of-low-carbon-steel: hunker.com, Retrieved 10 August, 2019

- Medium-carbon-steels, engineering: sciencedirect.com, Retrieved 28 February, 2019

- The-uses-for-medium-carbon-steel: hunker.com, Retrieved 8 March, 2019

- High-carbon-steel: instituteofmaking.org.uk, Retrieved 14 May, 2019

- High-carbon-steel-properties-uses-7596348: sciencing.com, Retrieved 1 March, 2019

- Stainless-Steel: madehow.com, Retrieved 12 July, 2019

- Austenitic-stainless-steel: corrosionpedia.com, Retrieved 20 January, 2019

- Martensitic-stainless-steel: corrosionpedia.com, Retrieved 22 May, 2019

- Martensitic-stainless-steels: ispatguru.com, Retrieved 1 February, 2019

- What-is-structural-steel: masteel.co.uk, Retrieved 14 June, 2019

- How-is-tool-steel-made: metalsupermarkets.com, Retrieved 3 April, 2019

Physical and Mechanical Properties of Steel

The physical properties of steel include high strength, resistance to corrosion and low weight. The mechanical properties of steel are its weldability, durability, ductility and malleability. The topics elaborated in this chapter will help in gaining a better perspective about these mechanical and physical properties of steel.

Physical Properties of Steel

Steel has a density of 7,850 kg/m³, making it 7.85 times as dense as water. Its melting point of 1,510 C is higher than that of most metals. In comparison, the melting point of bronze is 1,040 C, that of copper is 1,083 C, that of cast iron is 1,300 C, and that of nickel is 1,453 C. Tungsten, however, melts at a searing 3,410 C, which is not surprising since this element is used in light bulb filaments.

Steel's coefficient of linear expansion at 20 C, in μm per meter per degree Celsius, is 11.1, which makes is more resistant to changing size with changes in temperature than, for example, copper (16.7), tin (21.4) and lead (29.1).

Additives to Steel

Small amounts of other metals added to steel change its properties in ways favorable to certain industrial applications. For example, cobalt results in higher magnetic permeability and is used in magnets. Manganese adds strength and hardness, and the product is suitable for heavy-duty railway crossings. Molybdenum maintains its strength at high temperatures, so this additive is handy when making speed drill tips. Nickel and chromium resist corrosion and are usually added in the manufacture of steel surgical instruments.

The Base Metal: Iron

The major component of steel is iron, a metal that in its pure state is not much harder than copper. Omitting very extreme cases, iron in its solid state is, like all other metals, polycrystalline—that is, it consists of many crystals that join one another on their boundaries. A crystal is a well-ordered arrangement of atoms that can best be pictured as spheres touching one another. They are ordered in planes, called lattices, which penetrate one another in specific ways. For iron, the lattice arrangement can best be visualized by a unit cube with eight iron atoms at its corners. Important for the uniqueness of steel is the allotropy of iron—that is, its existence in two crystalline forms. In the body-centred cubic (bcc) arrangement, there is an additional iron atom in the centre of each cube. In the face-centred cubic (fcc) arrangement, there is one additional iron atom at the centre of each of the six faces of the unit cube. It is significant that the sides of the face-centred cube, or the distances between neighbouring lattices in the fcc arrangement, are about 25 percent larger than in the bcc arrangement; this means that there is more space in the fcc than in the bcc structure to keep foreign (i.e., alloying) atoms in solid solution.

Iron has its bcc allotropy below 912 °C (1,674 °F) and from 1,394 °C (2,541 °F) up to its melting point of 1,538 °C (2,800 °F). Referred to as ferrite, iron in its bcc formation is also called alpha iron in the lower temperature range and delta iron in the higher temperature zone. Between 912° and 1,394 °C iron is in its fcc order, which is called austenite or gamma iron. The allotropic behaviour of iron is retained with few exceptions in steel, even when the alloy contains considerable amounts of other elements.

There is also the term beta iron, which refers not to mechanical properties but rather to the strong magnetic characteristics of iron. Below 770 °C (1,420 °F), iron is ferromagnetic; the temperature above which it loses this property is often called the Curie point.

Effects of Carbon

In its pure form, iron is soft and generally not useful as an engineering material; the principal method of strengthening it and converting it into steel is by adding small amounts of carbon. In solid steel, carbon is generally found in two forms. Either it is in solid solution in austenite and ferrite or it is found as a carbide. The carbide form can be iron carbide (Fe_3C, known as cementite), or it can be a carbide of an alloying element such as titanium. (On the other hand, in gray iron, carbon appears as flakes or clusters of graphite, owing to the presence of silicon, which suppresses carbide formation.)

The effects of carbon are best illustrated by an iron-carbon equilibrium diagram. The A-B-C line represents the liquidus points (i.e., the temperatures at which molten iron begins to solidify), and the H-J-E-C line represents the solidus points (at which solidification is completed). The A-B-C line indicates that solidification temperatures decrease as the carbon content of an iron melt is increased. (This explains why gray iron, which contains more than 2 percent carbon, is processed at much lower temperatures than steel.) Molten steel containing, for example, a carbon content of 0.77 percent (shown by the vertical dashed line in the figure) begins to solidify at about 1,475 °C (2,660 °F) and is completely solid at about 1,400 °C (2,550 °F). From this point down, the iron crystals are all in an austenitic—i.e., fcc—arrangement and contain all of the carbon in solid solution. Cooling further, a dramatic change takes place at about 727 °C (1,341 °F) when the austenite crystals transform into a fine lamellar structure consisting of alternating platelets of ferrite and iron carbide. This microstructure is called pearlite, and the change is called the eutectoidic transformation. Pearlite has a diamond pyramid hardness (DPH) of approximately 200 kilograms-force per square millimetre (285,000 pounds per square inch), compared with a DPH of 70 kilograms-force per square millimetre for pure iron. Cooling steel with a lower carbon content (e.g., 0.25 percent) results in a microstructure containing about 50 percent pearlite and 50 percent ferrite; this is softer than pearlite, with a DPH of about 130. Steel with more than 0.77 percent carbon—for instance, 1.05 percent—contains in its microstructure pearlite and cementite; it is harder than pearlite and may have a DPH of 250.

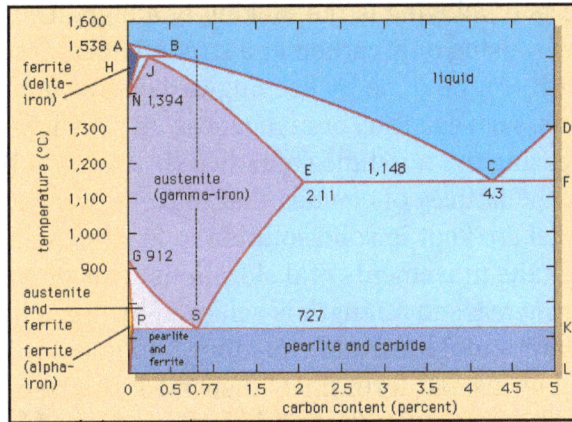

Iron-carbon equilibrium diagram.

Effects of Heat-treating

Adjusting the carbon content is the simplest way to change the mechanical properties of steel. Additional changes are made possible by heat-treating—for instance, by accelerating the rate of cooling through the austenite-to-ferrite transformation point, shown by the P-S-K line in the figure. (This transformation is also called the Ar1 transformation, r standing for refroidissement, or "cooling.") Increasing the cooling rate of pearlitic steel (0.77 percent carbon) to about 200 °C per minute generates a DPH of about 300, and cooling at 400 °C per minute raises the DPH to about 400. The reason for this increasing hardness is the formation of a finer pearlite and ferrite microstructure than can be obtained during slow cooling in ambient air. In principle, when steel cools quickly, there is less time for carbon atoms to move through the lattices and form larger carbides. Cooling even faster—for instance, by quenching the steel at about 1,000 °C per minute—results in a complete depression of carbide formation and forces the undercooled ferrite to hold a large amount of carbon atoms in solution for which it actually has no room. This generates a new microstructure, martensite. The DPH of martensite is about 1,000; it is the hardest and most brittle form of steel. Tempering martensitic steel—i.e., raising its temperature to a point such as 400 °C and holding it for a time—decreases the hardness and brittleness and produces a strong and tough steel. Quench-and-temper heat treatments are applied at many different cooling rates, holding times, and temperatures; they constitute a very important means of controlling steel's properties.

Effects of Alloying

A third way to change the properties of steel is by adding alloying elements other than carbon that produce characteristics not achievable in plain carbon steel. Each of the approximately 20 elements used for alloying steel has a distinct influence on microstructure and on the temperature, holding time, and cooling rates at which microstructures change. They alter the transformation points between ferrite and austenite, modify solution and diffusion rates, and compete with other elements in forming intermetallic compounds such as carbides and nitrides. There is a huge amount of empirical information on how alloying affects heat-treatment conditions, microstructures, and properties. In addition, there is a good theoretical understanding of principles, which, with the help of computers, enables engineers to predict the microstructures and properties of steel when alloying, hot-rolling, heat-treating, and cold-forming in any way.

A good example of the effects of alloying is the making of a high-strength steel with good weldability. This cannot be done by using only carbon as a strengthener, because carbon creates brittle zones around the weld, but it can be done by keeping carbon low and adding small amounts of other strengthening elements, such as nickel or manganese. In principle, the strengthening of metals is accomplished by increasing the resistance of lattice structures to the motion of dislocations. Dislocations are failures in the lattices of crystals that make it possible for metals to be formed. When elements such as nickel are kept in solid solution in ferrite, their atoms become embedded in the iron lattices and block the movements of dislocations. This phenomenon is called solution hardening. An even greater increase in strength is achieved by precipitation hardening, in which certain elements (e.g., titanium, niobium, and vanadium) do not stay in solid solution in ferrite during the cooling of steel but instead form finely dispersed, extremely small carbide or nitride crystals, which also effectively restrict the flow of dislocations. In addition, most of these strong carbide or nitride formers generate a small grain size, because their precipitates have a nucleation effect and slow down crystal growth during recrystallization of the cooling metal. Producing a small grain size is another method of strengthening steel, since grain boundaries also restrain the flow of dislocations.

Alloying elements have a strong influence on heat-treating, because they tend to slow the diffusion of atoms through the iron lattices and thereby delay the allotropic transformations. This means, for example, that the extremely hard martensite, which is normally produced by fast quenching, can be produced at lower cooling rates. This results in less internal stress and, most important, a deeper hardened zone in the workpiece. Improved hardenability is achieved by adding such elements as manganese, molybdenum, chromium, nickel, and boron. These alloying agents also permit tempering at higher temperatures, which generates better ductility at the same hardness and strength.

Mechanical Properties of Steel

The most important properties of steels which account for their widespread use are their mechanical properties. These properties include a combination of very high strength with the ability to bend rather than break. Different tests have been developed to describe the strength and ductility (a measure of bendability) of steels. A number of these tests which are used to describe the mechanical properties of steels are described below:

Tensile Testing

Tensile testing of steel is a kind of a testing done for the evaluation of the strength of steels. A length of the steel material, usually a round cylindrical rod, is pulled apart in a machine that applies a known force, F. The machine has grips which are attached to the ends of the cylindrical steel rod, and the force is applied parallel to the axis of the rod, as shown schematically in figure. As the force increases, the rod gets longer, and the change in length is represented as delta l (? l), where the symbol delta (?) means 'a change in' and the l refers to the original length of the rod. If a force of 50 kg is applied to two rods of the same steel material, where one is thin and the other thick then the thin rod elongate more. To compare their mechanical properties independent of rod diameter, the term 'stress' is used. Stress is simply the force divided by the cross-sectional area of the rod. When the same stress is applied to the thin and thick rods, they elongate the same amount, because the actual force applied to the thick rod is now larger than that applied to the thin rod by an amount proportional to its larger area. Because stress is force per area, its unit is Pascal (Pa) or N/sq m. However, the usual unit commonly used for describing the stress of the steel materials is mega-pascal (MPa) or N/sq mm (1 MPa = 1 N/sq mm).

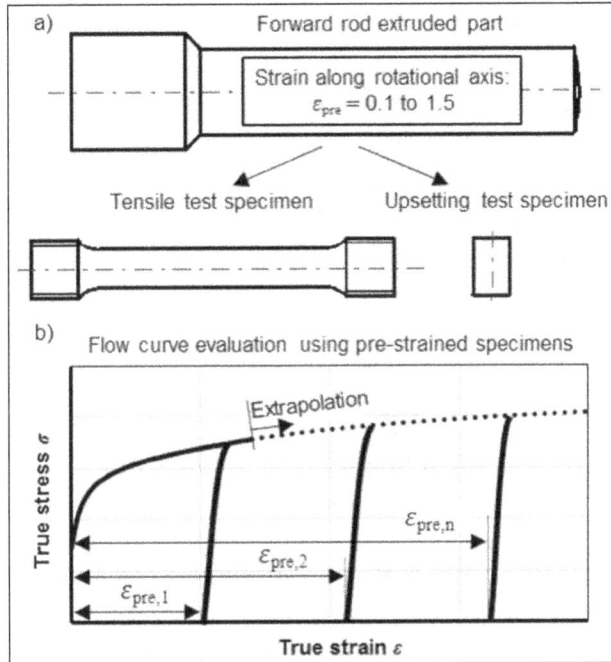

Stress- strain diagram for steels.

When a steel material is pulled along its axis, the applied force is called a tensile force, and the machine that applies the force is called a tensile testing machine. Figure shows a typical result obtained from a tensile test of a sample of steel material. The applied stress is plotted on the vertical axis while the change in length is plotted on the horizontal axis. It is usual to plot the fractional change in length, delta l/l (? l/l), as shown in figure. The fractional change in length is called the strain, and the diagram in figure is called a stress-strain diagram.

The stress-strain diagram is generally divided into regions, as shown in figure. These regions are the elastic region and the plastic region. As the stress on the steel sample increases, the sample elongates, and as long as the stress is not too high, release of the stress returns the steel sample to its original length. This is called elastic deformation. However, if the applied stress reaches a critical level, called the yield stress (YS), the steel material gives in and two things happen namely (i) the increase in stress needed to produce a given small increase in strain becomes lower, and (ii) on release of the stress, the sample is permanently elongated, as shown by the arrowed line A-B in figure. In this case, the sample is stressed to point A, and after releasing the stress, the sample is elongated from its original length by a percent (%) given as B × 100. As shown in the figure, the increase in stress needed to continue elongating the sample reaches a maximum value in the plastic region and then drops a little before the stress is able to break the sample in two pieces. This maximum stress value is normally called the ultimate tensile strength (UTS) or often just the tensile strength (TS).

The stress-strain diagram also provides an additional measure of the mechanical properties of the steel material. The ductility is the amount of elongation which occurs after the stress increases beyond the YS and before the sample breaks. This elongation is sometimes called permanent elongation, because it remains in the sample after breakage and can be measured easily. The permanent elongation in the sample in figure after the elastic strain is relaxed is given by point C. By simply multiplying the strain at C by 100, the '% elongation' of the steel material is obtained.

Figure also presents possible stress-strain diagrams for a ductile and a brittle steel material. The breakage strain is much larger for the ductile steel material, and the % elongation is much larger. The figure also shows that the diameter of the steel sample at the fracture surface of a brittle failure remains close to its original value, while that of the ductile failure is reduced. This reduction in diameter by plastic flow near the fracture surface is referred to as necking, and it develops in ductile steel material just before fracture. In addition to % elongation, ductility is often characterized by % reduction in area, which is simply the % by which the original cross-sectional area of the steel sample is reduced at the fractured neck.

The concept of ductile versus brittle behaviour is quite obvious and can be seen by hammering the exposed end of a sample fixed in a vice. In case of a brittle steel material, the sample breaks almost immediately, whereas in case of a ductile steel material, the sample is bent by the hammer blows and may not break even after bending 90 deg or more. Further if the sample is hammered back and forth, it can be noticed that after the material has been deformed a little, it becomes more difficult to deform further. This effect is called work hardening and can be understood from the stress-strain diagram in figure. The stress-strain diagram for the original (un-deformed) material is shown by the dotted line. If the original sample is deformed to point A, it has strain equivalent to point B when the stress is released. If this sample is retested, it can be notices that its YS has increased from the original value up to point A, as shown on the diagram. This increase in YS is a measure of the work hardening resulting from the original deformation (work) put into the steel material during the first deformation.

It can be seen that although the steel material is now stronger in the sense that it has to be stressed more before it undergoes plastic deformation, its UTS is not changed significantly, and it is less ductile, that is, its % elongation to failure has gone down. It is the general characteristic of the steel materials which shows that processes which increase their YS also decrease their ductility. This means that heavily work hardened steel materials often break easily. A familiar example of this behaviour is the frequently used method of breaking a steel wire by bending it back and forth many times.

The mechanical properties of steels are usually shown by listing values of their YS, TS, and % elongation. The data on the mechanical properties of normalized steels shows two general characteristics of steels namely (i) increasing the % carbon (C) in steel increases both YS and TS, and (ii) increasing % C drops the ductility (% elongation). Similarly a comparison of the as-rolled properties to the normalized properties shows the effects of work hardening. This comparison shows that YS increases at higher C levels, TS is changed only slightly, and % elongation drops a little.

Hardness Testing

A problem with the tensile testing is that the material is destroyed by the test. Another testing method which characterizes the strength of a steel material but does not destroy the material is the hardness test. This test is used widely because it is quick and can be applied to parts that can then be placed into service.

Over the years, several different, useful hardness testing methods have been developed, and the essential features of these testing methods are explained in figure. A hard material, called the indenter, is forced into the steel material surface with a fixed load (weight). The region of steel material located under the indenter point is deformed to strains which are well into the plastic region of figure, so that permanent deformation is generated, causing a crater (called an indent) to be left in the surface of the steel material. The hardness is then defined by some number which is proportional to the size of the indent. In some methods, the size of the indent is measured from its diameter, and in others, it is measured from the depth of the indent. There are various indentation hardness tests which are used for measuring the hardness of steel materials.

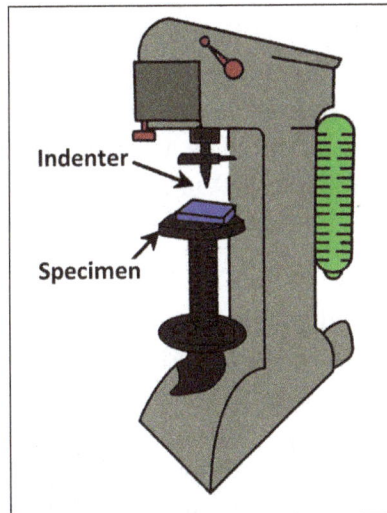

Essential features of hardness testing.

The Different Hardness Tests are Described Below:

- Rockwell hardness test – The Rockwell hardness test is probably the most widely used test. There are three most common scales for the Rockwell hardness test. Both the 'C' and 'A' scales use a conical diamond indenter, with the only difference being that the 'A' scale uses a lighter load. These tests provide a very quick and easy measurement, with the testing machine measuring the depth of penetration automatically and providing the hardness value either on a dial or, in modern testing machines, as a digital readout. The smaller load for the 'A' scale reduces the penetration depth and is often used to measure hardness on steel surfaces that have been case hardened with a thin, hard layer on the surface. The reported hardness values are referred to as RC or HRC and similarly RA or HRA. The indenter for the 'B' scale is a 1.6 mm hardened ball and this scale is sometimes used for soft steel materials.

- Brinell hardness test – It is a more reliable hardness test for soft-to-medium hard steels. It uses a larger ball made of either hardened steel or tungsten carbide, for indentation, and the diameter of the indent is measured. This test is a two step test in that the indent needs to be made first and then the indent diameter measured optically. The size of the indent is required to be between 3 mm to 6 mm. The Brinell hardness number (referred to usually as HB or BHN) is then determined either from a table or by using an equation. An advantage of the Brinell hardness test is that it measures hardness over a much larger area than the Rockwell hardness test, which is useful for materials with coarse microstructures, such as cast irons. However, this is also a disadvantage for application where the test piece is usually small, such as the tooth of a gear. Another disadvantage is that the test is not useful at the higher hardness ranges of steel materials. There is information available in standards which compares HRC values measured on steel with the corresponding equivalent hardnesses measured for steel by various other hardness tests.

- Vickers/Diamond pyramid hardness test – There is another test developed in England and generally used there in preference to the Rockwell test, called the Vickers hardness test. This test is similar to the Brinell test in that it is a two-step test measuring the diameter of the indent, but with a diamond indenter shaped in the form of a pyramid. The pyramid geometry produces an indent having a square shape, and the size is measured by the length of the diagonals of the square indent. The average diagonal length and the load used are put into a formula which calculates the Vickers hardness (HV) number, also called the diamond pyramid hardness (DPH) number. The Vickers test is the only one that applies at all hardness levels.

- Micro-hardness test (Diamond pyramid and Knoop) – There is a micro-hardness test available. The test uses very light loads and makes indent diameters small enough to fit into micro-sized regions. It uses a microscope that allows the indent to be positioned at desired locations on a microstructure, and the size of the indent is measured with the same microscope. There are two different indenters used in micro-hardness testing. They are the diamond pyramid indenter of the Vickers test and a special-shaped indenter called a Knoop indenter. Figure explains the use of the diamond pyramid (Vickers) hardness test on a ferrite-pearlite banded steel material. Indents are placed in a ferrite band and in a pearlite band. It can be seen that the indent is smaller in the pearlite band, indicating that the pearlite is harder than the ferrite. The weight used in this test is to be adjusted to 50 g to keep the indent small enough to fit inside the bands. The micro-hardness test uses the same equation as the Vickers hardness test to calculate hardness, HV, and the values found are HV = 260 in pearlite and HV = 211 in ferrite. Comparing this, these values correspond to HRC values of 24 in pearlite and 13.6 in ferrite. The second image in figure presents results from a similar test on ferrite-pearlite banded steel with higher sulphur (S), using the Knoop indenter. This steel sample contains significant amounts of sulphides due to higher S, as shown in the figure. The sulphides are ductile at the hot working temperature and become elongated during hot working. The sulphides make the steel machine more easily. It can be seen that the Knoop indenter is quite oblong in shape. This allows it to fit into thin layers more easily than a Vickers indent. Therefore, it is particularly useful for microstructures with thin layer morphologies. Hardness is characterized with its own set of Knoop hardness numbers, which can be related to HRC values in comparison tables.

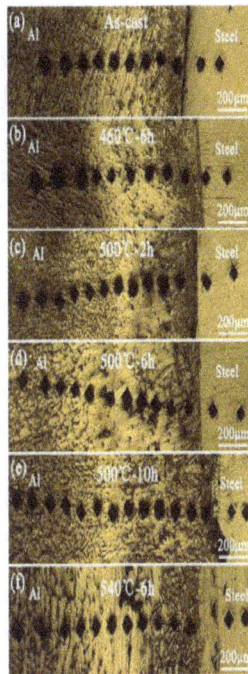

Micro-hardness testing of steels.

The indentation hardness numbers correlate well with the UTS of quenched and tempered steels. The correlation becomes less reliable at HRC values above 55, but the figure usually shown in tables shows the UTS values that are obtained by extrapolating the curve to the HRC values of 60 and 65. The results show that when a steel material is hardened into the range of HRC = 60 to 65, the corresponding tensile strengths are extremely high, 2450 N/sq mm and 2850 N/sq mm at HRC = 60 and 65, respectively.

While estimating the maximum hardness for the various microstructures in steels it can be seen that both ferrite and austenite are fairly soft. The hardness of pearlite is controlled by its spacing, and the values of 40 to 43 are obtained for pearlite with the finest spacing that can be obtained on cooling. Finer spacing can be obtained in pearlitic steels by mechanically deforming this pearlite to thin wire form and winding into cable. Such pearlitic steel cable is used in wire rope, which has TSs of 2200 N/sq mm. Quenched and tempered steel wire can be made to this strength level, but experience has found the pearlitic wire to be tougher, and for that reason, it is the wire of choice for such industrial applications as bridges and crane cables. Lower bainite has hardnesses approaching that of martensite and also finds industrial uses because of its slightly superior toughness to that of quenched and tempered steels. Martensite has the highest strength and hardness of all, and the hardness of fresh martensite depends on the % C in the steel. Fresh martensite is rarely used industrially because of its lack of toughness. It is tempered (heated to modest temperatures), which lowers its strength but increases its toughness. Measurement of hardness is a major method of control of the tempering process.

Notched Impact Testing

During World War II, the need for higher production rates of warships led to construction using welded steel plate rather than the standard riveted construction. The brittle failure in the welded plates of these ships produced spectacular failures, where the entire ship broke in half, with catastrophic results. Cracks began at local points in the welded joints and propagated around the ship, passing from plate to plate and causing failure by a brittle mode, that is, little to no plastic flow. This brittle behaviour was not detected by a sudden loss in ductility in the simple tensile test. These disasters led to a wide appreciation of the fact that ductility as measured by the tensile test is not a good measure of susceptibility to brittle behaviour in complex steel parts.

Ferritic steels, with their body-centered cubic (bcc) structure, have the disadvantage of breaking in a brittle fashion at low temperatures. This means, in terms of the ideas of the tensile test that the % elongation at failure is close to zero. As the temperature is lowered, there is a small temperature range over which the steels with bcc structure suddenly begin to fail in the brittle mode. An average temperature of the small range, called the 'ductile brittle transition temperature' (DBTT), is often chosen to characterize the temperature where the transition occurs. The simple tensile test detects this transition, but unfortunately, it detects DBTT values well below those that occur in complex steel parts.

The tensile test applies stress in only one direction while in complex steel parts, the applied stress acts in all three possible directions, a situation called a tri-axial stress state. The DBTT is raised by a tri-axial stress state. A tri-axial stress state develops at the base of a notch when a notched sample is broken in a tensile machine, and such tests are called notched tensile tests. However, it is more useful to break the sample with an impact test, where the load is applied much more rapidly than in a tensile machine, because the combination of the notch geometry and the high load rate produces values of DBTT close to the temperature where brittle failure begins to occur in complex steel parts.

The Charpy impact test, as shown schematically in figure, is most often used for the impact test. A V-notch is machined into a square bar and placed in the holder. The specimen is broken cleanly in two with a weighted hammer attached to the arm of a pendulum. The pendulum arm is raised to a specified height and then released. It swings down through the sample and swings up on the opposite side. By comparing the rise after breaking the sample to the rise with no sample present, the amount of energy absorbed by the specimen on breakage can be calculated. This energy, usually given in units of joules (J), is called the Charpy impact energy, or CVN energy. It can be seen that the 'Izod impact test' is similar to the Charpy impact test. The main difference in the Izod impact test is that the specimen is clamped at one end, and the pendulum strikes the opposite end.

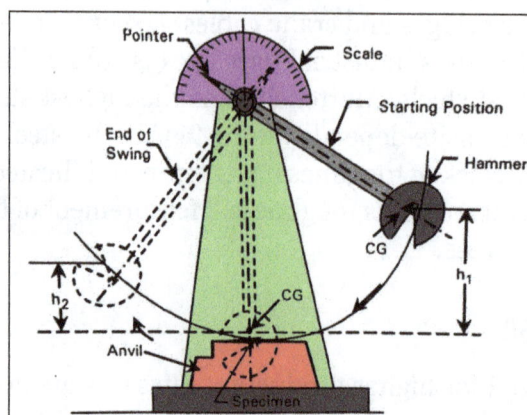

Charpy impact test.

Figure shows Charpy impact data on plain C steels which are slow cooled from the austenite region, so that the microstructures of these steels are mixtures of ferrite and pearlite. The figure shows that the Charpy impact energy is very sensitive to % C in the steels. At all compositions, there is a transition from ductile to brittle failure as the temperature drops, but at % C levels of 0.11 and below, the transition is much sharper and the CVN energy is considerably higher for the high temperature ductile mode of fracture. For steels, a CVN energy value of 20 J is often taken as the onset value for brittle failure. The dashed line in figure at 20 J shows that for this criterion, steels with % C above around 0.5 are considered brittle at room temperature. However, the DBTT predicted by the Charpy test is really only a guide to actual transition temperatures that occurs for a complex piece of steel used in the field. So, the data in figure do not mean that a steel having 0.60 % C will always fail by brittle fracture at room temperature.

It is common to refer to the CVN energy as a measure of the notch toughness, or simply the toughness of the steel. This measure of toughness is much more useful than ductility measurements, such as % elongation, for evaluating the potential of a steel material for brittle failure in service. Toughness and ductility have related but slightly different meanings, and it is high toughness that is desirable in steel parts. The toughness of steels can be improved by the factors namely (i) minimizing the C content, (ii) minimizing the grain size, (iii) eliminating inclusions, such as the sulphide stringers usually done by using steels with low impurity levels of S, P (phosphorus), and other elements, and (iv) using a microstructure of either quenched and tempered martensite or lower bainite rather than upper bainite or ferrite / pearlite.

There is a large difference in the nature of the fracture mechanism when a steel material fails in a brittle mode versus a ductile mode. In brittle failure, there is little to no plastic flow of the material prior to a crack forming and running through the steel, causing it to break into pieces. The steel seems to pull apart along the fracture surface. This sudden pulling apart occurs in one of two ways namely (i) the steel material separates along grain boundaries, or (ii) the grains of steel material are cleaved in half along certain planes of their crystal structure. These brittle fracture modes are called grain-boundary fracture and cleavage fracture and can be distinguished by looking at the fracture surface in either a low-power microscope or a scanning electron microscope (SEM).

When a steel material fails in a ductile mode, there is some plastic flow of the material prior to breakage occurring along the fracture surface. This flow causes very tiny voids to form within the steel grains. As the flow continues, these voids grow and coalesce until breakage occurs. The SEM micrographs at fairly high magnifications show remnants of the individual voids exposed on the fracture surface. The fracture is often called 'micro-void coalescence', for obvious reasons. If the steel contains small foreign particles, they are frequently be the site where the voids form first, and these particles are found at the bottom of the voids on one or the other of the mating fracture surfaces. Unfortunately, it is difficult to see the micro-voids on fracture surfaces with optical microscopes, because the depth of field is too shallow to image them at the high magnifications needed.

Grain-boundary fracture surfaces virtually always indicate that brittle failure has occurred, with little to no plastic flow before fracture. Cleavage fracture surfaces sometimes occur with significant plastic flow and, by themselves, do not indicate brittle failure. Micro-void coalescence fracture surfaces indicate that a ductile fracture occurred.

Fatigue Failure and Residual Stresses

A crude but quick way to break a steel rod in two is to make a small notch across the surface with a file or a cold chisel and then bend the rod at the notch. The rod generally breaks with a single bend or may be two. When the rod is bent away from the notch, a tensile stress develops at the base of the notch that is much higher than would be present if the notch are not there, and this leads to fracture starting at the notch. The small radius of curvature at the root of the notch causes a local rise in the stress there. The sharp radius is a stress concentrator or, alternately, a stress intensifier, and the smaller the radius, the higher the stress concentration. Small surface scratches, even those too small to see by eye, can produce stress concentrations during bending, which are localized at the root of the scratch. If sufficiently high, the concentrated stress can exceed the YS locally and produce tiny cracks that have the potential to lead to failure.

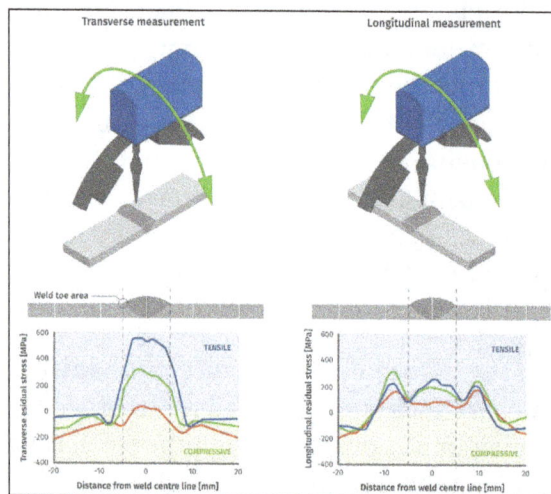

Compressive and tensile stresses on a rotating axis.

This awareness can help in understanding a type of failure that occurs in steels and is called fatigue failure. Consider an axle supporting a load heavy enough to cause it to bend down slightly at its centre point between the wheels. This bending causes the steel to be pulled apart at the location T and also to be pushed together at the location C. This means that the steel material undergoes a tensile stress at T and a compressive stress at C. Therefore, as a point on the axle surface rotates, it undergoes a cyclic stress: tension when it is down and compression when it is up. In a well designed axle, the maximum TS is usually well below the YS, and all the strain experienced in the steel material surface during rotation is in the elastic region of the stress-strain diagram, as shown at the bottom of figure.

However, if there is a very small scratch on the steel surface. If the stress is raised locally at the scratch root to a point above the YS, it can generate a crack there. Each time the axle rotates, the crack grows, and eventually, the crack becomes large enough to break the axle, a process known as fatigue failure. Fatigue failures occur in steel parts subjected to cyclic stresses, as often occurs in rotating machinery, valve and coil springs, or vibrating parts, such as aircraft wings.

When a steel bar is pulled on with a TS, the atoms of the iron grains are pulled away from each other, so that, on average, the distance between atoms increases. Similarly, if a compressive stress is applied, the average distance between atoms decreases a small amount. Hence, if the steel is stress free, there is some average distance between atoms. A steel rod lying on the bench generally is assumed to be stress free; that is, the atoms are spaced at their stress-free distance. Because the rod is just sitting there, this must be true, but only for the average distance between atoms over the whole rod. It may be that in parts of the rod, the atoms are sitting there with average distances between them greater than their stress-free distance, and in other parts, less than this value. If this is the case, the rod is said to contain residual stresses, some of them tensile and some of them compressive. Suppose, for example, that the outer millimeter of the rod is under compression. Then, below this outer 1 mm thick cylindrical region there must be some region where the atoms are under tension. In this case, there is residual stress on the surface that is compressive, with residual tension below. It is the surface where the scratches are that leads to fatigue failure. So, by producing residual compressive stresses on the surface, it is possible to reduce the crack growth during cyclic loading. As shown in figure, the local residual compressive stress at the surface causes the TS on the rotating axle to decrease, because the cyclic stress produced by rotation is added to the axle, starting at the value of the residual compressive surface stress. Thus, it is very desirable to have residual compressive stresses on steel parts.

There are several ways to produce residual compressive stresses in steels. These involve heat treating and/or mechanical deformation. The heat treating technique is called flame hardening. In this technique the surface of the steel heats up much faster than its interior since an intense flame is directed at the surface. The flame heating causes a layer of the steel at the surface to become austenite, while the interior remains ferrite + pearlite. On rapid cooling with a quench of some sort, the surface layer transforms to martensite, while the inner region remains ferrite + pearlite. Martensite has a lower density than the austenite from which it forms. Hence, the outer layer expands when martensite forms. The inner region resists this expansion, causing the atoms in the outer layer to be forced closer together than they want to be, and a residual compressive stress is formed in the outer martensitic layer. An alternate way of heating the surface layer much faster than the interior is by induction heating. A copper coil is placed at the surface of the steel, and a high- frequency current is passed through the coil. The magnetic field generated by this current induces currents in the steel surface, causing localized surface heating. This technique is widely used in industry. It is so effective that axles used in automobiles are routinely induction hardened, not for increased surface wear resistance but simply for improved fatigue life.

Residual compressive stresses can also be produced on heat treating, due simply to thermal expansion / contraction effects that occur on heating and cooling. If a round steel bar is quenched, the outer surface thermally contracts faster than the interior. The interior resists the contraction of the outer surface and pushes it into tension. If the tension is large enough to cause plastic flow in the outer surface, a stress reversal occurs on further cooling, and the result is a residual compressive stress on the outer surface.

As with flame hardening and induction hardening, the end result is a desirable residual compressive surface stress. However, if a phase transformation occurs on cooling, things become more complicated. So, for steels, the compressive residual surface stress is assured if the steel is not heated above the A1 temperature before quenching. Heating above A1 generates austenite, which complicates matters due to the formation of martensite on quenching. If martensite forms only in an outer surface layer on quenching, then a surface residual compressive stress is obtained.

However, if martensite develops all the way to the centre of the bar (a condition called through hardening), the result is a residual surface tensile stress, an undesirable situation.

There are many surface treatments that are given to steels. The two most common such treatments involve carburizing and nitriding. These treatments are usually done to produce hardened layers on the surface that reduce surface wear rates. These treatments also enhance the formation of residual compressive stresses, which is incidental benefit accompanying the increased surface hardness.

Residual surface compressive stresses can also be generated by mechanical means. Perhaps the most common technique employed in industry is shot peening. The technique involves bombarding the steel surface with small steel balls (shot). This action causes localized plastic flow at the surface, resulting in the desired residual surface compressive stress. Steel toughness is enhanced by surface residual compressive stresses. An example involves the high-strength pearlitic wire used in cables.

In general, pearlitic steel is not as tough as quenched and tempered martensitic steels. Yet, the high-strength pearlitic wire used in bridge and crane cables displays superior toughness compared to quenched and tempered martensitic wires of the same hardness. The improved toughness of the pearlitic wire is thought to arise from the drawing operation used to refine the pearlite spacing, which is necessary to achieve the high strength. Apparently, the wire-drawing operation produces residual surface compressive stresses that result in the improved toughness.

Weldability of Steels

There are several factors which control the weldability of carbon (C) and low alloy steels in electric arc welding. A good understanding of the chemical and physical phenomena which occurs in the weldments is necessary for the proper welding of the different steels. Operational parameters, thermal cycles, and metallurgical factors affecting the weld metal transformations and the susceptibility to hot and cold cracking are some of the factors which have marked influence on the weldability of steels. There are also some common tests which determine the weldability of steel.

The C and low alloy steels represents a large number of steels which differ in chemical composition, strength, heat treatment, corrosion resistance, and weldability. These steels can be categorized as (i) plain C steels, (ii) high strength low alloy (HSLA) steels, (iii) quenched and tempered (QT) steels, (iv) heat treatable low alloy (HTLA) steels, and (v) pre-coated steels.

To understand weldability of steels, it is necessary to have knowledge about the various weld regions.

Characteristic Features of Welds

Single Pass Weldments

In the case of a single pass bead, the weldment is generally divided into two main regions namely (i) the fusion zone, or weld metal, and (ii) the heat affected zone (HAZ) as shown in figure. Within the fusion zone, the peak temperature exceeds the melting point of the base steel, and the chemical composition of the weld metal depends on the choice of welding consumables, the base steel dilution ratio, and the operating conditions. Under conditions of rapid cooling and solidification of the weld metal, alloying and impurity elements segregate extensively to the centre of the inter-dendritic or inter-cellular regions and to the centre parts of the weld, resulting in significant local chemical in-homogeneities. Therefore, the transformation behaviour of the weld metal can be quite different from that of the base steel, even when the bulk chemical composition is not significantly changed by the welding process. The typical anisotropic nature of the solidified weld and structure is also shown in figure.

The chemical composition remains largely unchanged in the HAZ region since the highest temperature remains below the melting point of the base steel. However, considerable microstructural change takes place within the HAZ region during welding as a result of the extremely severe thermal cycles. The material immediately adjacent to the fusion zone is heated high into the austenitic temperature range. The micro-alloy precipitates which has developed in the previous stages of processing get generally dissolved and unpinning of austenite grain boundaries occurs with substantial growth of the grains, forming the coarse grain in the HAZ region. The average size of the austenite grains, which is a function of the highest temperature attained, decrease with increasing distance from the fusion zone. The cooling rate also varies from point to point in the HAZ region. It increases with increasing highest temperature at constant heat input and decreases with increasing heat input at constant highest temperature. Because of varying thermal conditions as a function of distance from the fusion line, the HAZ region is essentially composed of coarse grain zone (CGHAZ), fine grain zone (FGHAZ), inter-critical zone (ICHAZ), and subcritical zone (SCHAZ). The various HAZ regions of a single pass low C steel butt weld are shown in figure.

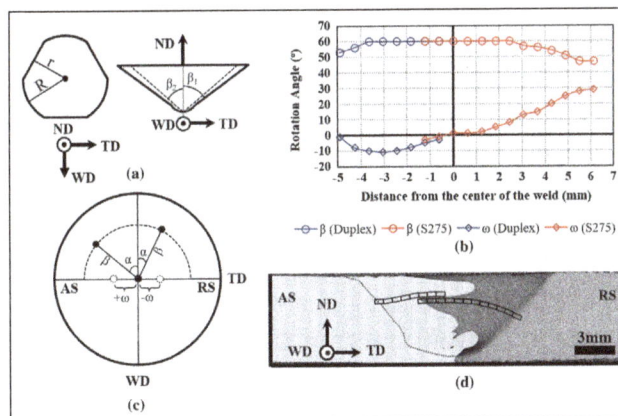

Weld structure and HAZ regions in a single pass low C steel but weld.

Multi Pass Weldments

The situation in the multi pass weldments is much more complex because of the presence of re-heated zones within the fusion zone. The partial refinement of the microstructure by subsequent weld passes increases the inhomogeneity of the various regions with respect to the microstructure and the mechanical properties. Re-austenitization and subcritical heating can have a profound effect on the subsequent structures and properties of the HAZ region. Toughness property deterioration is related to small regions of limited ductility and low cleavage resistance within the

CGHAZ which are identified as the localized brittle zone (LBZ). LBZs consist of unaltered CGHAZ, inter-critically reheated coarse grain (IRCG) HAZ, and sub-critically reheated coarse grain (SRCG) HAZ. LBZs can be aligned at an adjacent fusion line. The aligned LBZs offer short and easy paths for crack propagation. Fracture occurs along the fusion line.

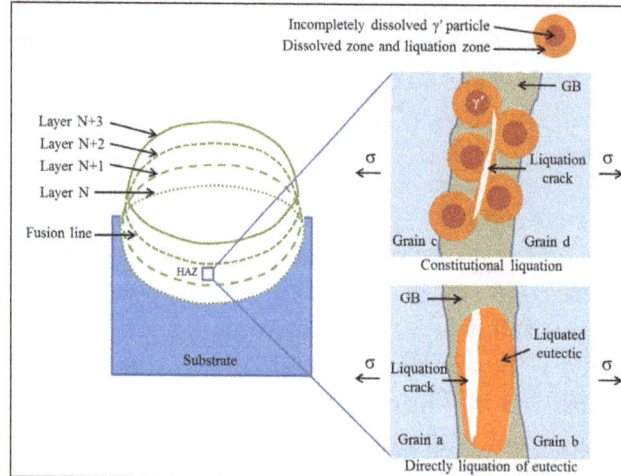

Overlapping of HAZ to form localized brittle zones aligned along the fusion line.

Metallurgical Factors Affecting Weldability

The metallurgical factors which affect the weldability are described below:

Hardenability

Hardenability in steels is generally used to indicate austenite stability with alloy additions. It is also being used as an indicator of weldability and as a guide for selection a material and welding process to avoid excessive hardness and cracking in the HAZ region. Steels with high hardness often contain a high volume fraction of martensite, which is exceedingly susceptible to cracking during the processing. Hardenability is also used to indicate the susceptibility of steel to hydrogen (H2) induced cracking. Traditionally, empirical equations have been developed experimentally to express weldability of steels. Carbon equivalent (CE) is one such terminology. It has been developed to estimate the cracking susceptibility of a steel during welding and to determine whether the steel needs pre-weld or/and post-weld heat treatments to avoid cracking. CE equations include the hardenability effect of the alloying elements by expressing the chemical composition of the steel as a sum of weighted alloy contents. Many CE terminologies with different coefficients for the alloying elements are being used. The equation for the CE as given by the 'International Institute of Welding (IIW)' is given below:

$$CE = C + Mn/6 + (Ni + Cu)/15 + (Cr + Mo + V)/5$$

In this equation the concentration of the alloying elements is given in weight percent. It can be seen in the equation that C is the element which has maximum effects on the weldability. Along with other chemical elements, C can affect the solidification temperature range, hot tear susceptibility, hardenability, and cold cracking behaviour of the steel weldment.

The application of CE terminologies is also empirical. As an example, the CE equation of IIW has been used effectively with usual medium C low alloy steels. Steels with lower CE values generally show good weldability. When the CE of the steel is less than 0.45 %, weld cracking is unlikely, and no heat treatment is needed. When the CE of the steel is between 0.45 % and 0.60 %, weld cracking is likely to take place, and preheating in the range of around 100 deg C to 400 deg C is generally suggested. When the CE of the steel is more than 0.60 %, there is a high probability for the weld to crack, and in such case both preheating and the post-weld heat treatments are needed to get a sound weld.

However, the CE equation of IIW does not accurately correlate with the microstructures and properties of low C micro-alloyed steels over extended alloy ranges. Thus, new terminologies based on solution thermodynamics and kinetic considerations have been developed to obtain better predictions of the alloy behaviour and weldability of low C low alloy steels. Complex interactive terms, rather than simple additive forms, are included in these equations. Equations with these nonlinear terms are more useful in predicting electric arc welding behaviour.

Several terminologies are also available for other steel groups with a wider range of alloying elements and with different prior heat treatments, H_2 contents, and weld hardnesses. One such equation (given below) includes fabrication conditions such as heat input, cooling rate, joint design and restraint conditions have also been proposed.

$$Ph = Pcm + 0.075 \log_{10} H + Rf / 40,000$$

In this equation Ph is the cracking susceptibility parameter, H is the concentration of H_2 (in parts per million, ppm), Rf is the restraint stress (in N/sq mm), and Pcm is as per the following equation.

$$Pcm = C + (Mn + Cu + Cr) / 20 + Si/30 + Ni/60 + Mo/15 + V/10 + 5 B$$

The thickness of the steel being welded can also be related to CE as a compensated carbon equivalent (CCE) which is CCE = CE + 0.00254 t where t is the thickness of the part in millimeters.

The above equations are valid only for specific ranges of chemical composition and welding conditions. However, despite the different forms and terms included in the predictive equations, the main objective remains that of estimating the weldability and cracking susceptibility of the steel.

Weld Metal Microstructure

Inherent in the welding process is the formation of a pool of liquid steel directly below a moving heat source. The shape of this liquid pool is determined by the flow of both heat and metal, with melting occurring ahead of the heat source and solidification occurring behind it. Heat input determines the volume of liquid steel and therefore the dilution and weld metal composition, as well as the thermal conditions under which solidification takes place. Also important to solidification is the crystalline growth rate, which is geometrically related to weld travel speed and weld pool shape. Thus, weld pool shape, weld metal composition, cooling rate, and growth rate are all the factors which are interrelated with heat input, which in turn affects the solidification microstructure and the tolerance of the weldment to hot cracking.

Incipient melting at base steel grain boundaries immediately adjacent to the fusion zone allows these grains to serve as seed crystals for epitaxial (an oriented overgrowth of crystalline material upon the surface of another crystal of different chemical composition but similar structure) grain growth during weld metal solidification. The continuous growth of the epitaxial grains results in large columnar grains whose boundaries provide easy paths for crack propagation. An elongated weld pool yields straight and broad columnar grains, which promote the formation of centerline cracking because of impurity segregation, mechanical entrapment of inclusions, and the shrinkage stresses which develop during solidification. Epitaxial columnar growth is particularly harmful in multi-pass welds where grains can extend continuously from one weld bead to another.

Hot tears originate near the liquid / solid interface when strains from solidification shrinkage and thermal contraction cause rupture of the liquid films of low melting point located at grain boundaries. The susceptibility of steel to hot tearing is related to its inability to accommodate strain through dendrite interlocking as well as the tendency of tears to backfill with the remaining liquid. The time interval during which liquid films can exist in relation to the rate of strain generation can also play a role in hot tear susceptibility. Steels can be hot tear sensitive depending on the amount of phosphorus (P) and sulphur (S) impurities they contain. C and nickel (Ni) are also known to influence hot cracking in steel welding.

When the solidified steel weld metal cools down, solid state transformation reactions can occur. As in solidification, the two main factors which determine the final microstructure are the chemical composition and thermal cycle of the weld metal. In maximum structural steels, weld metal solidifies as delta – ferrite. At the peritectic temperature, austenite forms from the reaction between liquid weld metal and delta – ferrite, and subsequent cooling leads to the formation of alpha – ferrite. During the austenite to ferrite transformation, proeutectoid ferrite forms first along the austenite grain boundaries. This is known as grain boundary ferrite. Subsequent to grain boundary ferrite formation, ferrite side plates develop in the form of long needlelike ferrite laths that protrude from the allotriomorphs. A coarse austenite grain size and low C content, in combination with a relatively high degree of super cooling, are found to promote ferrite side plate formation. These laths can be properly characterized by their length to width aspect ratios (usual values are more than 10:1).

As the temperature continues to drop, intra-granular acicular ferrite nucleates and grows in the form of short laths separated by high angle boundaries. The inclination between orientations of adjacent acicular ferrite laths is usually larger than 20 degrees. The random orientation of these laths provides good resistance to crack propagation. Acicular ferrite laths have aspect ratios ranging from 3:1 to 10:1.

During proeutectoid ferrite formation, C is rejected continuously from the ferrite phase, enriching the remaining austenite, which later transforms into a variety of constituents, such as martensite (both lath and twinned), bainite, pearlite, and retained austenite. Because of the acicular nature of the bainite laths, they can also be described by their aspect ratio, with values similar to those of Widmanstätten side-plates. Very frequently, however, bainite laths occur in the form of packets associated with grain boundaries.

HAZ Microstructure

In terms of microstructure, long bainite laths with alternate layers of connected martensite islands are generally found in the CGHAZ of HSLA steel weldments. Martensite islands (martensite retained austenite constituents) are formed because of the enrichment of C in austenite in the inter-critical zone. Coarse austenite grain size in the near fusion region of the HAZ can suppress high temperature transformation products in favour of martensite and bainite on cooling. Upper bainite has a relatively high transformation temperature and is stable relative to the thermal cycles subsequent to those of the first pass. Fluctuation of the chemical composition of the micro-alloying elements can also contribute to CE change and to the amount of hard martensite present in the CGHAZ.

In the case of FGHAZ, even though the peak temperature attained is above thermal cycle Ac3, it is still well below the temperature for grain coarsening. The smaller prior austenite grain size and subsequent ferrite transformation produce a refined microstructure having grains smaller than those of the base steel. The microstructure is similar to that of normalized steel, with considerable toughness.

In case of ICHAZ, only partial transformation takes place resulting in a mixture of austenite and ferrite at the peak temperature of the thermal cycle. Upon cooling, the austenite in a matrix of soft ferrite decomposes, and the final microstructure depends on the bulk and local composition of the alloying elements. The cooling rate is also an important factor in determining the amount of martensite and bainite in the ferrite matrix.

In the case of SCHAZ, no observable microstructural changes are observed. Some spheroidization of carbides can occur. Upon reheating by subsequent weld passes, precipitates or pre-precipitate clusters can form, reducing the toughness. Irregularly shaped particles can also coalesce and strain the surrounding matrix, further lowering the toughness.

During HAZ thermal cycles between Ac1 and Ac3, the austenite gets enriched with C, which, upon cooling, transforms to martensite islands. In the as-welded condition, this transformation affects the IRCG region more than the other reheated zone.

Effect of Chemical Composition

The presence of a certain phase in the final microstructure of a weldment is generally explained by means of a continuous cooling transformation (CCT) diagram, which is formed by two sets of curves namely (i) the percent transformation curves, and (ii) the cooling curves. The percent transformation curves define the regions of stability of the different phases. The cooling curves represent the actual thermal conditions that the weld experienced. The intersection of these two sets of curves determines the final microstructure of the different weld zones. Typical CCT diagram for an HSLA steel weld metal is given in figure.

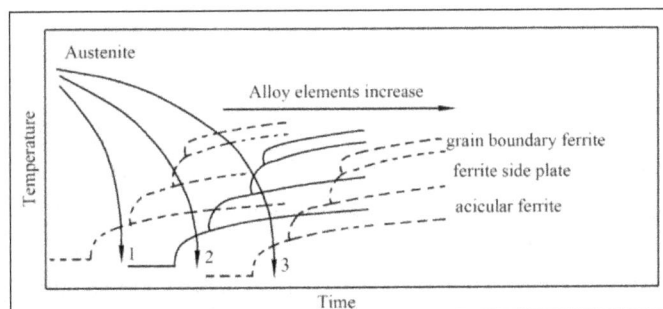

CCT diagram for an HSLA steel weld metal.

Hardenability elements, such as C, manganese (Mn), chromium (Cr), and molybdenum (Mo), suppress the start of austenite decomposition to lower temperatures. This is equivalent to pushing the transformation curves to the right side of the CCT diagram, resulting in a refined microstructure. Inclusion formers, such as oxygen (O_2) and S, accelerate the austenite to ferrite transformation by providing more nucleation sites for the reaction to initiate at higher temperatures. Faster cooling has the same effect as an increase in hardenability elements, while a slower cooling rate acts in the same direction as a decrease in hardenability agents or an increase in nucleation site providers. Since the cooling rate varies from point to point in the HAZ, the microstructure also changes accordingly, with martensite and bainite in regions close to the fusion line.

Pre-weld and Post-weld Heat Treatments

In the welding of C and low alloy steels, the final microstructure of the weldment is primarily determined by the cooling rate from the peak temperature. Since the alloy level in C and low alloy steels is low, the main physical properties of the steel are not affected. Thus, temperature gradient and heat input become the important parameters in weld metal microstructural evolution. A slower cooling rate decreases shrinkage stress, prevents excessive hardening, and allows time for H2 diffusion. Cooling rate is mainly important and is a function of the difference in temperature as well as the thermal conductivity of the steel.

During preheating, the initial temperature of the steel plate increases, decreasing the cooling rate and the amount of the hard phases, such as martensite and bainite, in the weld microstructure. For the welding of hardenable steels, it is important to determine the critical cooling rate (CCR) which the base steel can tolerate without cracking. The higher is the CE of the steel, the lower tis he critical or allowable cooling rate. The use of a low H2 welding electrode also becomes more important. Preheating is to be applied to adjust the cooling rate accordingly.

Weld Cracking

Maximum evidence indicates that a weld cracking failure mechanism is microstructure related. In the case of cold cracking, recent crack tip opening displacement (CTOD) results show that the reduction in toughness of HSLA weldments is related to the CGHAZ and that cracks normally propagate along or near the fusion line. The CGHAZ of micro-alloyed steel welds usually shows the highest hardness of the entire HAZ. The high C non tempered martensite in this region is the major cause of embrittlement. The amount of precipitates (carbides, nitrides, and carbo-nitrides) is found to be the highest in the regions next to the SCHAZ and the lowest at or next to the fusion line. As a result, there is a slight increase in micro-alloying element in solution in the CGHAZ, which increases the hardenability of this region.

H_2 Induced Cracking

The effect of H_2 on weld cracking is also a factor. Moisture pickup from the atmosphere that is incorporated into the liquid pool, either directly or via the welding consumables is the main source of H_2. The presence of H_2 increases the HAZ cracking susceptibility of high strength steel weldments. Also known as under-bead, cold, or delayed cracking, it is perhaps the most serious and least understood of all weld-cracking problems. It generally occurs at the temperature less than around 100 deg C either immediately upon cooling or after a period of several hours. The cracks can be both trans-granular and inter-crystalline in character. They largely follow prior austenite grain boundaries. The initiation of cold cracking is particularly associated with notches, such as the toe of the weld, or with in-homogeneities in microstructure which show sudden changes in hardness, such as slag inclusions, martensite/ferrite interfaces, or even grain boundaries.

Like most other crack growth phenomena, H2 induced cracking is accentuated in the presence of high restraint weld geometries and matrix hardening. Such cracking is associated with the combined presence of three factors namely (i) the presence of H2 in the steel (even very small amounts), (ii) a microstructure that is partly or wholly martensitic, and (iii) high residual stresses (generally as a result of thick material). If one or more of these conditions is absent or at a low level, H2 induced cracking does not take place. However, high cooling rates such as those found in manual processes further enhance the probability of weld HAZ cold cracking.

The tolerance of steels for H2 decreases with increasing C or alloy content. H2 induced cracking can be controlled by choosing a welding process or an electrode that produces little or no H2. Post-weld heat treatments can be used to decrease or eliminate the residual H2 or to produce a microstructure that is insensitive to H2 cracking. Lastly, welding procedures that result in low restraining stresses also reduce the risk of weld cracking.

Stress Relief Cracking

Stress relief cracking is also known as stress rupture cracking and reheat cracking. It is due to re-heating of the concern when welding quenched and tempered steel grades and heat resistant steels containing substantial amounts of carbide formers, such as Cr, Mo, and V. When weldments of these steels are heated more than around 500 deg C, inter-granular cracking along the prior austenite grain boundaries can take place in the CGHAZ. Stress relief cracking is thought to be closely related to the phenomenon of creep rupture. Furthermore, during reheating, the re-precipitation of carbides is likely to occur, further increasing the hardness. The precipitation of carbides during stress relaxation alters the delicate balance between resistance to grain boundary sliding and resistance to deformation within the coarse grains of the HAZ.

Certain procedures which can be used individually or in combination to decrease stress relief cracking in steels include the selection of a more appropriate weld joint design, weld location, and sequence of assembly to minimize restraint and stress concentrations. Selecting a filler metal which provides a weld metal that has significantly lower strength than that of the HAZ at the heat treating temperature is another way to minimize stress relief cracking. Peening each layer of weld metal to generate a surface compressive stress state that counteracts shrinkage stresses is also very effective.

Lamellar Cracking

Lamellar cracking is also known as lamellar tearing. It is characterized by a step like crack parallel to the rolling plane. The problem occurs particularly when making tee and corner joints in thick steel plates such that the fusion boundary of the weld runs parallel to the plate surface. High tensile stresses can develop perpendicular to the mid plane of the steel plate, as well as parallel to it. This tearing is usually associated with inclusions in the steel and progresses from one inclusion to another.

There is some indication that sensitivity to lamellar tearing is increased by the presence of H2 in the steel. Inclusions which contain low melting compounds, such as those of P and S, also increase the sensitivity of steel to lamellar tearing by wetting the prior austenite grain boundaries, since this make them too weak and fragile to withstand the thermal stresses during cooling. Some approaches which can minimize lamellar tearing are (i) changing the location and design of the welded joint to minimize through thickness strains, (ii) using a lower strength weld metal, (iii) reducing available H2, (iv) buttering the surface of the steel plate with weld metal prior to making the weld, (v) using preheat and inter pass temperatures of minimum 100 deg C, and (vi) using base plates with inclusion shape control.

Hot Cracking

Hot cracking is also known as solidification cracking. It occurs at higher temperatures and is generally located in the weld metal. Hot cracking also can be found in the HAZ, where it is known as liquation cracking. Hot cracking in weld deposits during cooling occurs predominately at the weld centerline or between columnar grains. The fracture path of a hot crack is inter-granular. The causes of hot cracking are well understood. The partition and rejection of alloying elements at columnar grain boundaries and ahead of the advancing solid/liquid interface produce significant segregation. The elements of segregation form low melting phases or eutectic structures to produce highly wetting films at grain boundaries. They weaken the structure to the extent that cracks form at the boundaries under the influence of the tensile residual stresses during cooling. Liquation cracking is also associated with grain-boundary segregation and is aggravated by the melting of these boundaries near the fusion line. These impurity weakened boundaries tend to rupture as the weld cools because of the high residual stresses.

Inclusions

Large amounts of P and S are added to some steels to provide in it the free machining characteristics. These steels have relatively poor weldability because of hot tearing in the weld metal caused by low melting compounds P and S at the grain boundaries. Iron oxide and iron sulfide inclusions, if present, are also harmful because of their solubility change with temperature and their tendency to precipitate at grain boundaries, contributing to low ductility, cracking, and porosity. Laminations which are flat separations or weaknesses which sometimes occur beneath and parallel to the surface of rolled steels have a slight tendency to open up if they extend to the weld joint.

Low C Steels

Low C steels are mostly used in structural applications. Steels with less than 0.15 % C can harden to 30 to 40 HRC (Rockwell hardness). Plain C steels containing less than 0.30 % C and 0.05 % S can be welded readily by maximum methods with little need for special measures to prevent weld cracking. The welding of sections that are more than 25 mm thick, particularly if the C content of the base steel exceeds 0.22 %, can need preheating to around 40 deg C and stress relieved at around 520 deg C to 670 deg C. For low C steels, a low alloy filler metal is generally suggested for meeting mechanical property requirements. The general procedure is to match the filler with the base steel in terms of strength or, for dissimilar welds, to match the lower strength steel. Frequently, however, higher strength weld metal can actually require a softer HAZ to undergo a relatively large amount of strain when the joint is subjected to deformation near room temperature. However, a low strength filler metal is not to be used indiscriminately as a remedy for cracking difficulties.

Medium C Steels

If steel containing around 0.5 % C is welded by a procedure normally used for low C steel, the HAZ is likely to be hard, low in toughness, and susceptible to cold cracking. Preheating the base steel can significantly reduce the rate at which the weld area cools, thus reducing the likelihood of martensite formation. Post-heating can further retard the cooling of the weld or can temper any martensite that might have formed. The suitable preheat temperature depends on the CE of the steel, the joint thicknesses, and the welding procedure. With a CE in the range of 0.45 % to 0.60 % range, a preheat temperature in the range of around 100 deg C is usually suggested. The minimum inter-pass temperature is to be the same as the preheat temperature. A low H2 welding procedure is required with these steels. Modifications in welding procedure, such as the use of a larger V-groove or of multiple passes, also decrease the cooling rate and the probability of weld cracking. Dilution can be minimized by depositing small weld beads or by using a welding procedure that provides shallow penetration. This is done to minimize C pickup from the base steel and the amount of hard transformation products in the fusion zone. Low heat input to limit dilution is also suggested for the first few layers in a multi-pass weld.

High C Steels

High C steels usually contain over 0.60 % C and show a very high elastic limit. They are frequently used in applications where high wear resistance is needed. These steels have high hardenability and sensitivity to cracking in both the weld metal and the HAZ region. A low H2 welding procedure is to be used for electric arc welding. Preheat and post-heat do not actually retard the formation of brittle high C martensite in the weld. However, preheating can minimize shrinkage stresses, and post-heating can temper the martensite that forms. Successful welding of high C steels needs the development of a specific welding procedure for each application. The composition, thickness, and configuration of the component parts need to be considered in process and consumable selections.

HSLA Steels

HSLA steels are designed to meet specific mechanical properties rather than a chemical composition. The alloy additions to HSLA steels strengthen the ferrite, promote hardenability, and help to control grain size. Weldability decreases as yield strength (YS) increases. For all practical purposes, welding these steels is the same as welding plain C steels which have similar CEs. Preheating may sometimes be needed, but post-heating is usually not required.

QT Steels

QT steels are furnished in the heat treated condition with YSs ranging from around 350 N/sq mm to 1000 N/sq mm, depending on the composition. The base steel is kept at less than 0.22 % C for good weldability. Preheating is to be used with caution when welding QT steels since it reduces the cooling rate of the weld HAZ. If the cooling rate is too slow, the re-austenitized zone adjacent to the weld metal can transform either to ferrite with regions of high C martensite, or to coarse bainite, of lower strength and toughness. A moderate preheat, however, can ensure against cracking, especially when the joint to be welded is thick and highly restrained. A post-weld stress-relief heat treatment is usually not required to prevent brittle fracture in weld joints in most QT steels.

HTLA Steels

The high hardness of HTLA steels requires that welding be conducted on materials in an annealed or over-tempered condition, followed by heat treatment to counter martensite formation and cold cracking. However, high preheating is often used with a low H2 process on these steels in a quenched and tempered condition. Preheating, or inter-pass heating, for both the weld metal and the HAZ are suggested. H2 control is also necessary to prevent weld cracking. Extremely clean vacuum melted steels are preferred for welding. Low P and S are needed in steels to reduce hot cracking. Segregation, which occurs because of the extended temperature range at which solidification takes place, reduces high-temperature strength and ductility. Fillers of lower C and alloy content are highly suggested. Preheat and inter-pass temperatures of 320 deg C or higher are very harsh environments for welders because of the physical discomfort and since an oxide layer forms at the weld joint. However, the cooling rate is to be controlled to allow the formation of a bainitic microstructure instead of the hard martensite. The bainitic microstructure can be heat treated afterward to restore the original mechanical properties of the structure. Specifications and procedures are to be followed thoroughly for these difficult-to-weld steels.

Pre-coated Steels

Thin steel plates and sheets are frequently pre-coated to protect them from oxidation and corrosion. The coatings usually used are aluminum (Al), zinc (Zn), and zinc rich primers. As anticipated, the coating originally at the weld region is destroyed during fusion welding, and the effectiveness of the coating adjacent to the weld is significantly reduced by the welding heat. In the case of aluminized steels, the formation of aluminum oxide can adversely affect the wetting and weld pool shape. The welding electrode and filler metals are to be selected carefully. A basic coating shielded metal arc welding (SMAW) electrode is suggested. For galvanized steels, weld cracking is generally attributed to inter-granular penetration by Zn. Zn dissolves considerably in iron to form an intermetallic compound at temperatures close to the melting temperature of Zn. Thus, molten Zn penetrates along the grain boundaries, leaving behind a brittle champagne fracture during cooling with the onset of a tensile stress state. Cracking occurs primarily at the throat region of a fillet weld, where shrinkage strain is more significant. The use of hot dipped coatings results in more severe cracking, while thin electro galvanized coatings are the least susceptible to cracking. Low silicon (Si) electrodes and rutile-based SMAW rods are both good for galvanized steel welding. Specific welding and setup procedures need to be followed, such as removing the Zn coating by an oxy-fuel process or by grinding, ensuring a large root opening, and using a slower welding speed to allow Zn vaporization and to prevent Zn entrapment in the weld metal.

Weldability Tests

Weldability tests are conducted to provide information on the service and performance of the welds. However, the data obtained in these tests can also be applied to the design of useful structures. Often these data are obtained from the same type of test specimens used in determining the base steel properties. Predicting the performance of structures from a laboratory type test is very complex because of the nature of the joint, which is far from homogeneous, metallurgically or chemically. Along with the base steel, the weld joint consists of the weld metal and the HAZ. Thus, a variety of properties are to be expected throughout the welded joint. Careful interpretation and application of the test results are needed.

There are presently many tests that evaluate not only the strength requirements of steel structures, but also the fracture characteristics and the effect of atmospheric conditions on early failure of the weldments.

Weld Tension Test

Several tension test specimens can be used for obtaining an accurate assessment of the strength and ductility of welds. Both the transverse weld specimens and longitudinal weld specimens are used. In the all weld metal test, base steel dilution need to be minimized if the test is to be representative of the weld metal. However, the resulting properties may not be easy to translate into those properties achievable from welds made in an actual weld joint.

Interpreting test results for the transverse butt weld test is complicated by the different strengths and ductilities normally found in the various regions of the joint. The primary information gained from the test is the ultimate tensile strength (UTS). YS and elongation requirements are usually not specified.

Tests of HAZ properties which are unaffected by the presence of either base steel or weld metal are not easy to conduct since it is practically impossible to obtain specimens made up entirely of the HAZ. In addition, the HAZ is composed of various regions, each with its own distinct properties. Simulated HAZ specimens that are generated and tested using a Gleeble thermo-mechanical testing system can be used to provide a more accurate assessment of the tensile properties of this region.

Bend Test

Different types of bend tests are used to evaluate the ductility and soundness of welded joints. Bend test results are expressed in various terms, such as percent elongation, minimum bend radius prior to failure, go/no-go for specific test conditions, and angle of bend prior to failure. Various specimen designs, both notched and un-notched, and testing techniques have been used. Nowadays, un-notched specimens can be used in quality control tests, while notched specimens may be used to predict in-service behaviour. However, most notched bend tests are used for research purposes and are not in common industrial use. Transverse bend tests are useful since they quite frequently reveal the presence of defects which are not detected in tension tests. However, the transverse specimen suffers from the same weakness as the transverse weld tension test specimen in that non-uniform properties along the length of the specimen can cause non-uniform bending, although this is often compensated for by the use of a wrap-around bend fixture.

Hardness Testing

Hardness testing can be used to complement information gained through tension or bend tests by providing information about the metallurgical changes caused by welding. Routine methods for the hardness testing of metals are well established. In C and low alloy steels, the hardness near the fusion line in the HAZ may be much higher than in the base steel because of the formation of martensite. In the HAZ areas where the temperature is low, the hardness may be lower than in the base steel because of tempering effects.

The Drop Weight Test

The drop weight test design is based on service failures resulting from brittle fracture initiation at a small flaw located in a region of high stress. The drop weight test can be considered a limited deflection bend test that uses a crack starter to introduce a running crack in the specimen. The specimen is a steel bar on which a brittle crack starter weld is deposited. This overlay cracks when the bar is deflected by the drop weight. A series of test is performed at different temperatures to determine the testing temperature below which the crack will propagate to the edges of the specimen. This critical temperature is also called the nil ductility temperature (NDT), defined as the highest temperature at which the propagating crack reaches the edge of the specimen. Therefore, the drop weight test is also known as the NDT test.

The Charpy V-notch (CVN) Impact Test

The Charpy V-notch impact test is the most popular technique for evaluating the impact properties of welds. The energy absorbed by a sample at fracture determines the toughness of the specimen. In this test, specimens at different temperatures are broken using a pendulum hammer. A typical plot of CVN results for a C and low alloy steel shows that there is a transition from low energy to high energy fracture over a narrow temperature range. This is associated with a change from trans-crystalline to ductile fracture. Therefore, material quality can be defined in terms of this transition temperature.

In the CVN impact testing of welds, the notch is typically located at the weld centerline. For CVN testing of the HAZ, the notch is more typically introduced at the CGHAZ. However, because precise location of a notch is never simple in the HAZ, simulated weld samples are used instead.

The Crack Tip Opening Displacement Test

The crack tip opening displacement (CTOD) test measures toughness, primarily for elastic-plastic conditions. In CTOD tests, the clip gauge opening at the onset of fracture is measured and used to calculate the crack opening displacement at the crack tip. The critical value of CTOD at fracture is a critical strain parameter that is analogous to the critical stress-intensity parameter. The CTOD test provides a useful method of determining the critical flaw size.

The CTOD test is very sensitive to changes in sample thickness, hardness, and strength, and it is difficult to obtain valid results in practical specimen thicknesses. The application of fracture mechanics to the prevention of catastrophic failure in weldments is, however, complicated by the nature of the weldment. In addition to their metallurgical heterogeneity, weldments often contain high residual stresses. Therefore, it is inadequate to fracture test the base steel and assume that the critical crack length thus determined is valid when the base steel is made into a weldment. The fracture toughness criterion needs to be determined for the base steel, the HAZ, and the weld metal. By first determining the zone with the lowest toughness value, it is then possible to evaluate a more realistic critical flaw size. However, the plane-strain fracture toughness tests are preproduction or pilot plant type tests that provide a rational means for designs and for estimating the effects of new designs, materials, or fabrication practices on the fracture-safe performance of structures.

Other popular tests include compact tension (CT) and wedge opening load (WOL) tests, which are normally used in the evaluation of structural weldments.

Stress Corrosion Cracking Test

The presence of corrosive atmospheres in a steel weldment can accelerate the initiation of a crack. Generally, the higher the strength of the steel, the more susceptible it becomes to stress corrosion cracking. The C and low alloy steels are not generally exposed to severely corrosive atmospheres, but rather to the normal atmosphere, moisture, hydrocarbons, fertilizers, and soils. However, welding can lower corrosive resistance by the introduction of (i) compositional differences which promote galvanic attack between weld metal, HAZ, and base steel when the joint in immersed in a conducting liquid, (ii) residual stresses that can cause stress corrosion cracking, and (iii) sur-face flaws that can act as sites for stress corrosion cracking. Stress corrosion cracking is normally delayed cracking, with longer time to failure at lower stresses. Maximum stress corrosion tests are fairly long in terms of time because of the slow crack initiation that occurs in un-notched test specimen. However, it has been found that the long initiation period can be eliminated by testing pre-cracked samples.

Fabrication Weldability Tests

There are various types of tests for determining the susceptibility of the weld joint to different types of cracking during fabrication. These tests are (i) restraint tests, (ii) externally loaded tests, (iii) under-bead cracking tests, and (iv) lamellar tearing tests.

The Lehigh Restraint Test

The Lehigh restraint test is particularly useful for quantitatively rating the crack susceptibility of a weld metal as affected by electrode variables. This test provides a means of imposing a control-lable severity of restraint on the root bead that is deposited in a butt weld groove with dimensions suitable to the application. Slots are cut in the sides and ends of a steel plate prior to welding. By changing the length of the slots, the degree of plate restraint on the weld is varied without sig-nificantly changing the cooling rate of the weld. Therefore, a critical restraint for cracking can be determined for given welding conditions. This sample is also useful for H_2 cracking.

The Varestraint Test

The Varestraint test determines the susceptibility of the welded joint to hot cracking. The test uti-lizes external loading to impose controlled plastic deformation in a steel plate while a weld bead is being deposited on the long axis of the steel plate. The specimen is mounted as a cantilever beam, and a pneumatically driven yoke is positioned to force the test piece downward when the welding arc reaches a predetermined position. By the choice of the radius to which the steel plate is bent, the severity of deformation causing cracking can be determined. Strain from 0 % to 4 % can be chosen according to the susceptibility of the joint to hot cracking. When the bending moment is applied transverse to the weld axis, the test is termed trans-varestraint test. A spot Varestraint test can also be conducted by keeping the arc stationary with bending applied at the moment the arc is extinguished.

The Controlled Thermal Severity Test

The controlled thermal severity test is designed to measure the cracking sensitivity of steels under cooling rates controlled by the thickness of the steel plates and the number of paths available for dissipating the welding heat. It is conducted with a steel plate bolted and anchor welded to a second steel plate in a position to provide two fillet (lap) welds. The fillet located at the plate edges has two paths of heat flow. The lap weld located near the middle of the bottom plate has three paths of heat flow, thus inducing faster cooling. The fillet welds are made first and allowed to cool, followed by the lap welds. After a holding time of 72 hours at room temperature, the degree of cracking is determined by measuring the crack length on metallographic specimens.

A number of other tests have been developed that contain welds in a circular configuration. The circular patch test has probably the most severe testing conditions. The two varieties are the Navy circular patch restraint test and the segmented circular patch restraint level. Cracking is detected by visual, radiographic, and liquid penetrant inspection. The cracking susceptibility of a material is measured as the total crack length and expressed as a percentage of the weld length. These tests can be used to determine both hot and cold cracking in the weld metal and the HAZ. Depending on the results, a go / no-go criteria are established for weld qualification.

Durability of Steels

Corrosion Resistance

- Properly designed and constructed steel structures provide long-term durability.

- Laboratory and field exposure tests on steel structures built to industry standard practices demonstrate excellent service life.

- Building codes and industry standards require that steel structures be designed to tolerate corrosion or be protected against corrosion where corrosion may impair strength or serviceability.

- Barrier coatings (such as paint) are readily available to coat the steel surface and isolate it from water and oxygen. Without water and oxygen, the steel cannot corrode.

- When further protection is needed, zinc coatings (such as galvanized) are available to provide sacrificial protection as well as barrier protection. When the base metal of a zinc-coated steel is exposed, such as at a cut or scratch, the steel is cathodically protected by the sacrificial corrosion of the zinc coating adjacent to the steel.

- When there are severe exposure conditions, such as industrial atmospheres and marine atmospheres, higher-performance coatings are available.

- For fabricated structural steel, HSS and open web steel joists, a variety of shop and field-applied paint and zinc coatings are available.

- For sheet steel, a variety of hot-dip galvanized coatings applied by the steel mill and/or paint coatings applied by a coil coating line are economical and readily available. These coatings are applied to the sheet steel prior to the coil being shipped to the manufacturer for roll forming of the finished product.

- Zinc-coated steel, which is standard for cold-formed steel framing, will last far beyond the life of a building when properly installed and insulated, and is especially appropriate in the high-demand structural configurations of mid-rise construction.

- Industry guidance is readily available from the steel construction associations. Following are some suggested resources:

  ○ AISC Design Guide 3: Serviceability Design Considerations for Steel Buildings.

  ○ CFSEI Technical Note D001-13: Durability of Cold-Formed Steel Framing Members.

  ○ MBMA Condensation Fact Sheet, 2009.

Mold Resistance

- Steel framing is resistant to mold since it is inorganic and does not provide a food source for mold to grow.

- Steel framing can help resist the onset and growth of mold since its framing members are dimensionally straight and connected mechanically (screwed vs. nailed), offering a tight envelope with no nail pops or drywall cracks (e.g., where the roof meets the walls). This makes the building structure stronger and more resilient.

- Ventilation is efficiently built into a steel-framed design, and energy efficiency is maintained or increased due to steel's inorganic properties.

- Moisture does not get into steel studs, substantially eliminating the expansion and contraction of construction materials around windows and doors, where leaks can occur.

Vermin Resistance

- Termites cause more damage to structures than fire, floods, and storms combined. Of particular concern is the Formosan termite, one of the most destructive termite species in the world. Originally limited to Hawaii, it is now well established throughout the southern United States around the Gulf Coast and spreading rapidly.

- Steel framing is not vulnerable to termites since it is inorganic and does not provide a food source for them.

- Cold-formed steel is one of the recognized methods for compliance with the termite-resistant construction requirements of the International Residential Code.

- There is no need for annual termite treatments with steel.

- Cold-formed steel provides a healthy building with no off-gassing from chemical termite treatments or pressure-treated lumber.

- Termite damage is rarely covered by insurance. Building with steel allows owners to avoid costly problems later.

Dimensional Stability

- When considering framing systems, particularly for mid-rise structures, the dimensional stability of the framing materials must be given careful consideration before and during the design and construction process.

- Unintended structural movement can have expensive and potentially disastrous consequences on structural, mechanical, and finish systems.

- When using materials that shrink or swell with changes in moisture content and changes in relative humidity, it is important to consider a variety of implications whether the material is the only framing material employed or is used in combination with steel or other materials that do not exhibit the same changes in the presence of moisture.

- Dimensional stability concerns are magnified when materials are used in a mid-rise building. Although a material might be used successfully in low-rise buildings, the same construction practices cannot be assumed to be adequate for taller structures.

- The best way to avoid dimensional stability problems is to build with a dimensionally-stable material such as steel.

- Steel structures provide long-term, consistent performance.

- Steel does not expand or contract with moisture content.

- Steel does not warp, split, crack or creep.

- Steel is isotropic, meaning it has the same dimensional properties in all directions. Since there is no "grain," the strength of steel is the same up and down, side-to-side, and in all loading directions.

- When they get wet, both wood and brick will swell. When they dry out and cure, concrete and concrete block will shrink and form shrinkage cracks.

- Using steel not only solves issues with structural movement due to changes in moisture content or humidity, but eliminates or greatly reduces other moisture related issues such as rot and mold.

- Wood is particularly prone to dimensional instability:

  ◦ If wood is considered for the primary framing material, or even for components of the structure such as exposed wood beams or timber frame trusses, the shrinkage of the wood and the associated cost of the special detailing required must be considered.

  ◦ In addition, with the increased emphasis on energy conservation, the long-term effects of shrinkage on the building envelope and building energy and maintenance costs must also be considered.

○ The cost of repairing cracks in framing and finishes, as well as painting, caulking, sealing, and termite protection should all be considered.

Ductility of Steels

Ductility is the measure of a material's ability to plastically deform without fracturing when placed under a tensile stress that exceeds its yield strength. High ductility indicates that a material will be more apt to deform and not break whereas low ductility indicates that a material is brittle and will fracture before deforming much under a tensile load. Ductility depends largely on a material's chemical composition, a material's crystalline structure, and the temperature at which the ductility is being measured.

Ductility is not the same as malleability. Ductility is a measure of material deformation under a tensile stress, whereas malleability is a measure of material deformation under compressive stress. A material does not necessarily have to have both high ductility and high malleability. It could have high malleability and low ductility.

Metals that have high ductility include gold, platinum, silver and iron. Low ductility metals include tungsten and steels with high amounts of carbon. Polymers are usually ductile; however there are brittle polymers available. Ceramics are typically brittle.

The ductility of a material will change as its temperature is changed. Metals have a ductile to brittle transition temperature. For polymers this is called a glass transition temperature. The exact temperature is different for different materials, but once it is reached, ductility is vastly reduced and the material becomes brittle. The ductile to brittle transition temperature or glass transition temperature of a material is an important consideration for materials subject to extreme cold.

The ductility of steel is influenced by the carbide distribution which can vary from spheroidal particles to lamellar pearlitic cementite. Comparing spheroidal cementite with sulphides of similar morphology, the carbide particles are stronger and do not crack or exhibit decohesion at small strains, with the result that a spheroidised steel can withstand substantial deformation before voids are nucleated and so exhibits good ductility. The strain needed for void nucleation decreases with increasing volume fraction of carbide and so can be linked to the carbon content of the steel.

Pearlitic cementite does not crack at small strains, but the critical strain for void nucleation is lower than for spheroidised carbides. Another factor which reduces the overall ductility of pearlitic steels is the fact that once a single lamella cracks, the crack is transmitted over much of a pearlite colony leading to well-defined cracks in the pearlite regions. The result is that the normal ductile dimpled fractures are obtained with fractured pearlite at the base of the dimples.

The effects of second phases on the ductility of steel are summarised in figure, where the sulphides are shown to have a more pronounced effect than either carbide distribution. This arises because, in the case of the sulphide inclusions, voids nucleate at a very early stage of the deformation process. The secondary effect of the particle shape both for carbides and sulphides is also indicated.

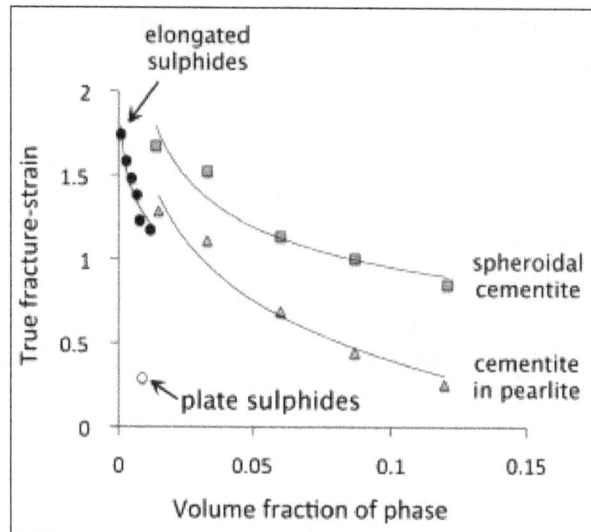

Effect of second-phase particles on the ductility of steel. Adapted from Gladman et al.

Malleability of Steels

Malleability is a physical property of metals that defines the ability to be hammered, pressed or rolled into thin sheets without breaking. In other words, it is the property of a metal to deform under compression onto a different form.

A metal's malleability can be measured by how much pressure (compressive stress) it can withstand without breaking. Differences in malleability among different metals are due to variances in their crystal structures.

Compression stress forces atoms to roll over each other into new positions without breaking their metallic bond. When a large amount of stress is put on a malleable metal, the atoms roll over each other, permanently staying in their new position.

Examples of malleable metals are:

- Gold
- Silver
- Iron
- Aluminum

- Copper
- Tin
- Indium
- Lithium

Examples of products demonstrating malleability include gold leaf, lithium foil, and indium shot.

Malleability and Hardness

The crystal structure of harder metals, such as antimony and bismuth, makes it more difficult to press atoms into new positions without breaking. This is because the rows of atoms in the metal don't line-up.

In other words, more grain boundaries exist and metals tend to fracture at grain boundaries. Grain boundaries are areas where atoms are not as strongly connected. Therefore, the more grain boundaries a metal has, the harder, more brittle and less malleable it will be.

Malleability versus Ductility

While malleability is the property of a metal deforming under compression, ductility is the property of a metal allowing it to stretch without damage.

Copper is an example of a metal that has both good ductility (it can be stretched into wires) and good malleability (it can also be rolled into sheets).

While most malleable metals are also ductile, the two properties can be exclusive. Lead and tin, for example, are malleable and ductile when they are cold but become increasingly brittle when temperatures start rising towards their melting points.

Most metals, however, become more malleable when heated. This is due to the effect that temperature has on the crystal grains within metals.

Controlling Crystal Grains through Temperature

Temperature has a direct effect on the behavior of atoms, and in most metals heat results in atoms having a more regular arrangement. This reduces the number of grain boundaries, thereby, making the metal softer or more malleable.

An example of temperature's effect on metals can be seen with zinc, which is a brittle metal below 300°F (149°C). Yet when heated above this temperature, zinc can become so malleable it can be rolled into sheets.

In contrast to the effect of heat treatment, cold working—a process that involves rolling, drawing, or pressing causing plastic deformation a cold metal—tends to result in smaller grains, making the metal harder.

Beyond temperature, alloying is another common method of controlling grain sizes to make metals more workable. Brass, an alloy of copper and zinc, is harder than both individual metals because its grain structure is more resistant to compression stress attempting to forces the rows of atoms from shifting into new positions.

References

- Chemical-physical-properties-steel-5548364: sciencing.com, Retrieved 10 July, 2019

- Steel, technology: britannica.com, Retrieved 15 February, 2019

- Mechanical-properties-of-steels: ispatguru.com, Retrieved 17 April, 2019

- Weldability-of-steels: ispatguru.com, Retrieved 19 August, 2019

- Durability, why-choose-steel: buildusingsteel.org, Retrieved 29 July, 2019

- Ductility, chemistry: sciencedirect.com, Retrieved 18 March, 2019

- Malleability-2340002: thebalance.com, Retrieved 11 January, 2019

Steelmaking Process

The process of producing steel from iron ore or scrap is called steelmaking. It is broadly divided into two steps, namely, primary steelmaking and secondary steelmaking. This chapter has been carefully written to provide an easy understanding of these facets of the steelmaking process.

Manufacturing steel is the process to produce steel from Iron ore. In this process, contaminants of source iron such as nitrogen, silicon, phosphorus, sulfur and excess carbon are removed and different types of metallic or non-metallic elements such as molybdenum, nickel, copper, carbon, titanium and vanadium are added to produce different types of steels.

Iron Processing

Iron processing is use of a smelting process to turn the ore into a form from which products can be fashioned.

Iron (Fe) is a relatively dense metal with a silvery white appearance and distinctive magnetic properties. It constitutes 5 percent by weight of the Earth's crust, and it is the fourth most abundant element after oxygen, silicon, and aluminum. It melts at a temperature of 1,538 °C (2,800 °F).

Iron is allotropic—that is, it exists in different forms. Its crystal structure is either body-centred cubic (bcc) or face-centred cubic (fcc), depending on the temperature. In both crystallographic modifications, the basic configuration is a cube with iron atoms located at the corners. There is an extra atom in the centre of each cube in the bcc modification and in the centre of each face in the fcc. At room temperature, pure iron has a bcc structure referred to as alpha-ferrite; this persists until the temperature is raised to 912 °C (1,674 °F), when it transforms into an fcc arrangement known as austenite. With further heating, austenite remains until the temperature reaches 1,394 °C (2,541 °F), at which point the bcc structure reappears. This form of iron, called delta-ferrite, remains until the melting point is reached.

The pure metal is malleable and can be easily shaped by hammering, but apart from specialized electrical applications it is rarely used without adding other elements to improve its properties. Mostly it appears in iron-carbon alloys such as steels, which contain between 0.003 and about 2 percent carbon (the majority lying in the range of 0.01 to 1.2 percent), and cast irons with 2 to 4 percent carbon. At the carbon contents typical of steels, iron carbide (Fe_3C), also known as cementite, is formed; this leads to the formation of pearlite, which in a microscope can be seen to consist of alternate laths of alpha-ferrite and cementite. Cementite is harder and stronger than ferrite but is much less malleable, so that vastly differing mechanical properties are obtained by varying the amount of carbon. At the higher carbon contents typical of cast irons, carbon may separate out as either cementite or graphite, depending on the manufacturing conditions. Again, a wide range of properties is obtained. This versatility of iron-carbon alloys leads to their widespread use in engineering and explains why iron is by far the most important of all the industrial metals.

The primary objective of iron making is to release iron from chemical combination with oxygen, and, since the blast furnace is much the most efficient process, it receives the most attention here. Alternative methods known as direct reduction are used in over a score of countries, but less than 5 percent of iron is made this way. A third group of iron-making techniques classed as smelting-reduction is still in its infancy.

The Blast Furnace

Basically, the blast furnace is a countercurrent heat and oxygen exchanger in which rising combustion gas loses most of its heat on the way up, leaving the furnace at a temperature of about 200 °C (390 °F), while descending iron oxides are wholly converted to metallic iron. Process control and productivity improvements all follow from a consideration of these fundamental features. For example, the most important advance of the 20th century has been a switch from the use of randomly sized ore to evenly sized sinter and pellet charges. The main benefit is that the charge descends regularly, without sticking, because the narrowing of the range of particle sizes makes the gas flow more evenly, enhancing contact with the descending solids. (Even so, it is impossible to eliminate size variations completely; at the very least, some breakdown occurs between the sinter plant or coke ovens and the furnace.)

Structure

The furnace itself is a tall, vertical shaft that consists of a steel shell with a refractory lining of firebrick and graphite. Five sections can be identified. At the bottom is a parallel-sided hearth where liquid metal and slag collect, and this is surmounted by an inverted truncated cone known as the bosh. Air is blown into the furnace through tuyeres, water-cooled nozzles made of copper and mounted at the top of the hearth close to its junction with the bosh. A short vertical section called the bosh parallel, or the barrel, connects the bosh to the truncated upright cone that is the stack. Finally, the fifth and topmost section, through which the charge enters the furnace, is the throat. The lining in the bosh and hearth, where the highest temperatures occur, is usually made of carbon bricks, which are manufactured by pressing and baking a mixture of coke, anthracite, and pitch. Carbon is more resistant to the corrosive action of molten iron and slag than are the aluminosilicate firebricks used for the remainder of the lining. Firebrick quality is measured by the alumina (Al_2O_3) content, so that bricks containing 63 percent alumina are used in the bosh parallel, while 45 percent alumina is adequate for the stack.

Until recently, all blast furnaces used the double-bell system to introduce the charge into the stack. This equipment consists of two cones, called bells, each of which can be closed to provide a gastight seal. In operation, material is first deposited on the upper, smaller bell, which is then lowered a short distance to allow the charge to fall onto the larger bell. Next, the small bell is closed, and the large bell is lowered to allow the charge to drop into the furnace. In this way, gas is prevented from escaping into the atmosphere. Because it is difficult to distribute the burden evenly over the furnace cross section with this system, and because the abrasive action of the charge causes the bells to wear so that gas leakage eventually occurs, more and more furnaces are equipped with a bell-less top, in which the rate of material flow from each hopper is controlled by an adjustable gate and delivery to the stack is through a rotating chute whose angle of inclination can be altered. This arrangement gives good control of burden distribution, since successive portions of the charge can be placed in the furnace as rings of differing diameter. The charging pattern that gives the best furnace performance can then be found easily.

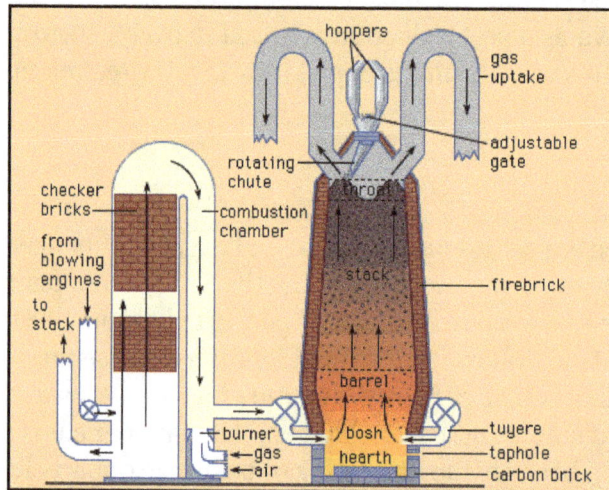

Schematic diagram of modern blast furnace (right) and hot-blast stove (left).

The general principles upon which blast-furnace design is based are as follows. Cold charge (mainly ore and coke), entering at the top of the stack, increases in temperature as it descends, so that it expands. For this reason the stack diameter must increase to let the charge move down freely, and typically the stack wall is displaced outward at an angle of 6° to 7° to the vertical. Eventually, melting of iron and slag takes place, and the voids between the solids are filled with liquid so that there is an apparent decrease in volume. This requires a smaller diameter, and the bosh wall therefore slopes inward and makes an angle to the vertical in the range of 6° to 9°. Over the years, the internal lines of the furnace that give it its characteristic shape have undergone a series of evolutionary changes, but the major alteration has been an increase in girth so that the ratio of height to bosh parallel has been progressively reduced as furnaces have become bigger.

For many years, the accepted method of building a furnace was to use the steel shell to give the structure rigidity and to support the stack with steel columns at regular intervals around the furnace. With very large furnaces, however, the mass is too great, so that a different construction must be used in which four large columns are joined to a box girder surrounding the furnace at a level near the top of the stack. The steel shell still takes most of the mass of the stack, but the furnace top is supported independently.

Operation

Solid charge is raised to the top of the furnace either in hydraulically operated skips or by the use of conveyor belts. Air blown into the furnace through the tuyeres is preheated to a temperature between 900° and 1,350 °C (1,650° and 2,450 °F) in hot-blast stoves, and in some cases it is enriched with up to 25 percent oxygen. The main product, molten pig iron (also called hot metal or blast-furnace iron), is tapped from the bottom of the furnace at regular intervals. Productivity is measured by dividing the output by the internal working volume of the furnace; 2 to 2.5 tons per cubic metre (125 to 150 pounds per cubic foot) can be obtained every 24 hours from furnaces with working volumes of 4,000 cubic metres (140,000 cubic feet).

Two by-products, slag and gas, are also formed. Slag leaves the furnace by the same taphole as the iron (upon which it floats), and its composition generally lies in the range of 30–40 percent silica (SiO_2), 5–15 percent alumina (Al_2O_3), 35–45 percent lime (CaO), and 5–15 percent magnesia (MgO). The gas exiting at the top of the furnace is composed mainly of carbon monoxide (CO), carbon dioxide (CO_2), and nitrogen (N_2); a typical composition would be 23 percent CO, 22 percent CO_2, 3 percent water, and 49 percent N_2. Its net combustion energy is roughly one-tenth that of methane. After the dust has been removed, this gas, together with some coke-oven gas, is burned in hot-blast stoves to heat the air blown in through the tuyeres. Hot-blast stoves are in effect temporary heat-storage devices consisting of a combustion chamber and a checkerwork of firebricks that absorb heat during the combustion period. When the stove is hot enough, combustion is stopped and cold air is blown through in the reverse direction, so that the checkerwork surrenders its heat to the air, which then travels to the furnace and enters via the tuyeres. Each furnace has three or four stoves to ensure a continuous supply of hot blast.

Chemistry

The internal workings of a blast furnace used to be something of a mystery, but iron-making chemistry is now well established. Coke burns in oxygen present in the air blast in a combustion reaction taking place near the bottom of the furnace immediately in front of the tuyeres:

$$2C + O_2 \rightarrow 2CO + \text{heat}$$

The heat generated by the reaction is carried upward by the rising gases and transferred to the descending charge. The CO in the gas then reacts with iron oxide in the stack, producing metallic iron and CO_2:

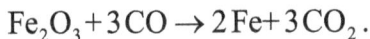

$$Fe_2O_3 + 3CO \rightarrow 2Fe + 3CO_2.$$

Not all the oxygen originally present in the ore is removed like this; some remaining oxide reacts directly with carbon at the higher temperatures encountered in the bosh:

$$FeO + C \rightarrow Fe + CO.$$

Softening and melting of the ore takes place here, droplets of metal and slag forming and trickling down through a layer of coke to collect on the hearth.

The conditions that cause the chemical reduction of iron oxides to occur also affect other oxides. All the phosphorus pentoxide (P_2O_5) and some of the silica and manganous oxide (MnO) are reduced, while phosphorus, silicon, and manganese all dissolve in the hot metal together with some carbon from the coke.

Direct Reduction (DR)

This is any process in which iron is extracted from ore at a temperature below the melting points of the materials involved. Gangue remains in the spongelike product, known as direct-reduced iron, or DRI, and must be removed in a subsequent steelmaking process. Only high-grade ores and pellets made from superconcentrates (66 percent iron) are therefore really suitable for DR iron making.

Direct reduction is used mostly in special circumstances, often linked to cheap supplies of natural gas. Several processes are based on the use of a slightly inclined rotating kiln to which ore, coal, and recycled material are charged at the upper end, with heat supplied by an oil or gas burner. Results are modest, however, compared to gas-based processes, many of which are conducted in shaft furnaces. In the most successful of these, known as the Midrex (after its developer, a division of the Midland-Ross Corporation), a gas reformer converts methane (CH_4) to a mixture of carbon monoxide and hydrogen (H_2) and feeds these gases to the top half of a small shaft furnace. There descending pellets are chemically reduced at a temperature of 850 °C (1,550 °F). The metallized charge is cooled in the bottom half of the shaft before being discharged.

Smelting Reduction

The scarcity of coking coals for blast-furnace use and the high cost of coke ovens are two reasons for the emergence of this other alternative iron-making process. Smelting reduction employs two units: in the first, iron ore is heated and reduced by gases exiting from the second unit, which is a smelter-gasifier supplied with coal and oxygen. The partially reduced ore is then smelted in the second unit, and liquid iron is produced. Smelting-reduction technologyenables a wide range of coals to be used for iron making.

Primary Steelmaking

Principles

In principle, steelmaking is a melting, purifying, and alloying process carried out at approximately 1,600 °C (2,900 °F) in molten conditions. Various chemical reactions are initiated, either in sequence or simultaneously, in order to arrive at specified chemical compositions and temperatures. Indeed, many of the reactions interfere with one another, requiring the use of process models to help in analyzing options, optimizing competing reactions, and designing efficient commercial practices.

Raw Materials

The major iron-bearing raw materials for steelmaking are blast-furnace iron, steel scrap, and direct-reduced iron (DRI). Liquid blast-furnace iron typically contains 3.8 to 4.5 percent carbon (C), 0.4 to 1.2 percent silicon (Si), 0.6 to 1.2 percent manganese (Mn), up to 0.2 percent phosphorus (P), and 0.04 percent sulfur (S). Its temperature is usually 1,400° to 1,500 °C (2,550° to 2,700 °F). The phosphorus content depends on the ore used, since phosphorus is not removed in the blast-furnace process, whereas sulfur is usually picked up during iron making from coke and other fuels. DRI is reduced from iron ore in the solid state by carbon monoxide(CO) and hydrogen (H_2). It frequently contains about 3 percent unreduced iron ore and 4 percent gangue, depending on the ore used. It is normally shipped in briquettes and charged into the steelmaking furnace like scrap. Steel scrap is metallic iron containing residuals, such as copper, tin, and chromium, that vary with its origin. Of the three major steelmaking processes—basic oxygen, open hearth, and electric arc—the first two, with few exceptions, use liquid blast-furnace iron and scrap as raw material and the latter uses a solid charge of scrap and DRI.

Oxidation Reactions

The most important chemical reactions carried out on these materials (especially on blast-furnace iron) are the oxidation of carbon to carbon monoxide, silicon to silica, manganese to manganous oxide, and phosphorus to phosphate, as follows:

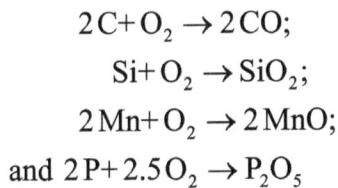

$$2C + O_2 \rightarrow 2CO;$$
$$Si + O_2 \rightarrow SiO_2;$$
$$2Mn + O_2 \rightarrow 2MnO;$$
$$\text{and } 2P + 2.5O_2 \rightarrow P_2O_5$$

Unfortunately, iron is also lost in this series of reactions, as it is oxidized to ferrous oxide:

$$2Fe + O_2 \rightarrow 2FeO.$$

The FeO, absorbed into the liquid slag, then acts as an oxidizer itself, as in the following reactions:

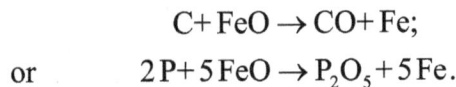

$$C + FeO \rightarrow CO + Fe;$$
$$\text{or} \qquad 2P + 5FeO \rightarrow P_2O_5 + 5Fe.$$

In the open-hearth furnace, oxidation also takes place when gases containing carbon dioxide(CO_2) contact the melt and react as follows:

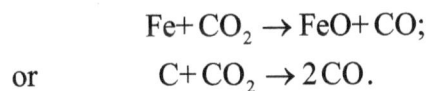

$$Fe + CO_2 \rightarrow FeO + CO;$$
$$\text{or} \qquad C + CO_2 \rightarrow 2CO.$$

The Slag

The products of the above reactions, the oxides silica, manganese oxide, phosphate, and ferrous oxide, together with burnt lime (calcium oxide; CaO) added as flux, form the slag. Burnt lime has by itself a high melting point of 2,570 °C (4,660 °F) and is therefore solid at steelmaking temperatures, but when it is mixed with the other oxides, they all melt together at lower temperatures and thus form the slag. A basic slag contains approximately 55 percent CaO, 15 percent SiO_2, 5 percent MnO, 18 percent FeO, and other oxides plus sulfides and phosphates. The basicity of a slag is often simply expressed by the ratio of CaO to SiO_2, with CaO being the basic and SiO_2 the acidic component. Usually, a basicity above 3.5 provides good absorption and holding capacity for calcium phosphates and calcium sulfides.

Removing Sulfur

The majority of sulfur, present as ferrous sulfide (FeS), is removed from the melt not by oxidation but by the conversion of calcium oxide to calcium sulfide:

$$FeS + CaO \rightarrow CaS + FeO.$$

According to this equation, desulfurization is successful only when using a slag with plenty of calcium oxide—in other words, with a high basicity. A low iron oxide content is also essential, since oxygen and sulfur compete to combine with the calcium. For this reason, many steel plants desulfurize blast-furnace iron before it is refined into steel, since at that stage it contains practically no dissolved oxygen, owing to its high silicon and carbon content. Nevertheless, sulfur is often introduced by scrap and flux during steelmaking, so that, in order to meet low sulfur specifications (for example, less than 0.008 percent), it is necessary to desulfurize the steel as well.

Removing Carbon

A very important chemical reaction during steelmaking is the oxidation of carbon. Its gaseous product, carbon monoxide, goes into the off-gas, but, before it does that, it generates the carbon monoxide boil, a phenomenon common to all steelmaking processes and very important for mixing. Mixing enhances chemical reactions, purges hydrogen and nitrogen, and improves heat transfer. Adjusting the carbon content is important, but it is often oxidized below specified levels, so that carbon powder must be injected to raise the carbon again.

Removing Oxygen

As the carbon level is lowered in liquid steel, the level of dissolved oxygen theoretically increases according to the relationship $\%C \times \%O = 0.0025$. This means that, for instance, a steel with 0.1 percent carbon, at equilibrium, contains about 0.025 percent, or 250 parts per million, dissolved oxygen. The level of dissolved oxygen in liquid steel must be lowered because oxygen reacts with carbon during solidification and forms carbon monoxide and blowholes in the cast. This reaction can start earlier, too, resulting in a dangerous carbon monoxide boil in the ladle. In addition, a high oxygen level creates many oxide inclusions that are harmful for most steel products. Therefore, usually at the end of steelmaking during the tapping stage, liquid steel is deoxidized by adding aluminum or silicon. Both elements are strong oxide formers and react with dissolved oxygen to form alumina (Al_2O_3) or silica. These float to the surface of the steel, where they are absorbed by the slag. The upward movement of these inclusions is often slow because they are small (*e.g.*, 0.05 millimetre), and combinations of various deoxidizers are sometimes used to form larger inclusions that float more readily. In addition, stirring the melt with argon or an electromagnetic field often serves to give them a lift.

Alloying

Deoxidation is also important before alloying steel with easy oxidizable metals such as chromium, titanium, and vanadium, in order to minimize losses and improve process control. Metals that do not oxidize readily, such as nickel, cobalt, molybdenum, and copper, can be added in the furnace to take advantage of high heating rates. In fact, alloying always has thermal effects on steelmaking—for example, the use of energy to heat and melt the alloying agents, or the heat of reaction or solution when they combine with other elements. Fortunately, there exists a large amount of empirical data, obtained from thousands of thermodynamic experiments, that, when supported by theoretical principles, allows steelmakers to predict such temperature changes.

Most alloys are added in the form of ferroalloys, which are iron-based alloys that are cheaper to produce than the pure metals. Many different grades are available. For example, ferrosilicon is supplied with levels of 50, 75, and 90 percent silicon and with varying levels of carbon and other additions.

Removing Hydrogen and Nitrogen

Also important for steelmaking is the absorption and removal of the two gases hydrogen and nitrogen. Hydrogen can enter liquid steel from moist air, damp refractories, and wet flux and alloy additions. It causes brittleness of solidified steel—especially in large pieces, such as heavy forgings, that do not permit the gas to diffuse to the surface. Hydrogen can also form blowholes in castings. Nitrogen does not move into and out of liquid steel as easily as hydrogen, but it is well absorbed by liquid steel in the high-temperature zones of an electric arc or oxygen jet, where nitrogen molecules (N_2) are broken up into atoms (N). Like hydrogen, nitrogen substantially decreases the ductility of steel.

Refractory Liner

Basic steelmaking takes place in containers lined with basic refractories. These may be bricks or ram material made of highly stable oxides, such as magnesite, alumina, or the double oxides chrome-magnesite and dolomite. It is desirable that the refractories not participate in the steelmaking reactions, but unfortunately they do erode and corrode. Refractory bricks are produced in all shapes and grades by a highly specialized industry.

Testing

Testing and sampling are an important part of liquid steelmaking. They are carried out by mechanized and often automated facilities, which immerse lances that are equipped with sensors for rapid computation of temperature and dissolved carbon, oxygen, and hydrogen. Test lances also take samples for analysis in laboratories. All results are usually fed automatically into a process-control computer.

Basic Oxygen Steelmaking

The most commonly applied process for steel-making is the integrated steel-making process via the Blast Furnace – Basic Oxygen Furnace. In the basic oxygen furnace, the iron is combined with varying amounts of steel scrap (less than 30%) and small amounts of flux. A lance is introduced in the vessel and blows 99% pure oxygen causing a temperature rise to 1700°C. The scrap melts, impurities are oxidised, and the carbon content is reduced by 90%, resulting in liquid steel.

The Charge

When oxygen contacts blast-furnace iron, a great amount of heat is released by the ensuing exothermic reactions, especially the oxidation of silicon to silica, so that using only blast-furnace iron would result in a liquid steel temperature too high for casting. Therefore, before the hot metal is added, a specific amount of scrap is charged into the furnace. Melting this scrap consumes about 340 kilocalories per kilogram, effectively cooling the process. A typical BOP charge, therefore, consists of about 75 percent liquid iron and 25 percent scrap. This requires a reliable supply of low-cost iron with a uniform chemical composition, which is attainable only by keeping the operating condition of a blast furnace as constant as possible; this in turn requires a consistent iron consumer. There are also certain iron properties—for example, the silicon and sulfur content—that are selected to optimize the blast furnace and BOF operations and to produce steel at minimal cost. Such interdependence requires that blast furnaces and BOFs work within a well-integrated operating system.

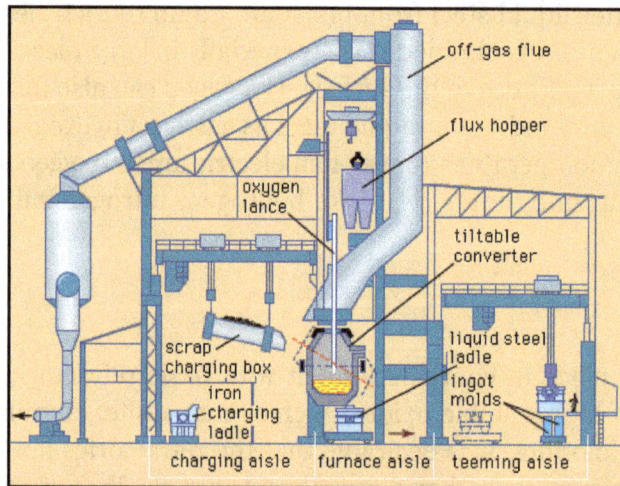

A basic oxygen furnace shop.

The Furnace

The basic oxygen converter is a cylindrical vessel with an open cone on top. For the largest converters, those that make 360-ton heats, the shell is about 8 metres in diameter and 11 metres high. The shells are built of heavy steel plates and sit in a trunnion ring so that the converter may be rotated for charging, testing, tapping, and slag-off. The lining, normally made of magnesite bricks, has different thickness and brick quality in certain zones, depending on the wear at each location. Total lining thickness of large converters exceeds one metre. The taphole is in the upper zone of the converter, right under the cone.

Oxygen lances are large, multiwall tubes that, on large converters, are about 300 millimetres in diameter and 21 metres long. Their tips have three to five nozzles, directed slightly outward, which produce the supersonic jets of oxygen. Proper water cooling of these lances is crucial. Special lance cranes move the lance up and down and adjust its distance from the steel bath. The lances last for about 150 heats before their tips have to be replaced.

BOFs are equipped with huge off-gas systems in order to avoid gas leakage into the shop and to ensure proper cleaning of the gases before they are discharged into the atmosphere. Off-gas emerges from the converter mouth at about 1,650 °C (3,000 °F). It consists of about 90 percent carbon monoxide and 10 percent carbon dioxide, and it also contains ferrous oxide dust, which forms in the high-temperature zone of the oxygen jet. Two off-gas systems are in use: the full combustion and the suppressed combustion.

In the full-combustion system, off-gas is burned above the mouth of the converter with excess air, and both physical and chemical heat are utilized in a boiler or hot-water system incorporated in the hood and vertical offtakes. A large venturi scrubber or electrostatic precipitator then cleans the cooled off-gas. During the blow of a large converter, about 10,000 cubic metres (350,000 cubic feet) of off-gas is moved per minute through full-combustion apparatus by exhaust fans, and about 0.7 kilogram of iron oxide dust is collected per ton of steel.

In the other system, the suppressed-combustion system, a ring-shaped hood is lowered onto the converter mouth before the blow, keeping air away from the hot off-gases. This means that they are not burned and that their chemical heating value of about 3,000 kilocalories per cubic metre is preserved. The gas is cleaned, collected in gas holders, and used at other locations. Though this system is more complicated, it is much smaller, because off-gases are cooler and there is less to be handled and processed.

BOFs are housed in huge buildings sometimes 80 metres high to accommodate the long lance, the off-gas system, and gravity-type feeding equipment. Heavy cranes, long conveyor belts, and railroad tracks assure prompt supply of raw material to the converters and fast removal of liquid steel and slag from the BOF.

The Process

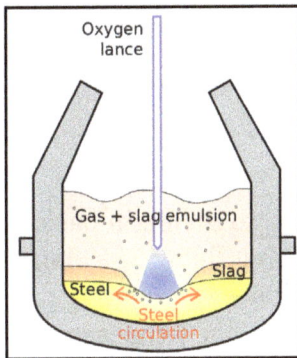

Principle of a LD converter.

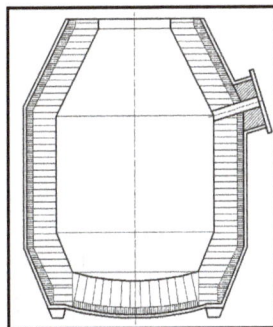

Cross-section of a basic oxygen furnace.

The outside of a basic oxygen steelmaking plant at the Scunthorpe steel works.

Basic oxygen steelmaking is a primary steelmaking process for converting molten pig iron into steel by blowing oxygen through a lance over the molten pig iron inside the converter. Exothermic heat is generated by the oxidation reactions during blowing.

The basic oxygen steel-making process is as follows:

Molten pig iron (sometimes referred to as "hot metal") from a blast furnace is poured into a large refractory-lined container called a ladle.

- The metal in the ladle is sent directly for basic oxygen steelmaking or to a pretreatment stage. High purity oxygen at a pressure of 700–1,000 kilopascals (100–150 psi) is introduced at supersonic speed onto the surface of the iron bath through a water-cooled lance, which is suspended in the vessel and kept a few feet above the bath. Pretreatment of the blast furnace hot metal is done externally to reduce sulphur, silicon, and phosphorus before charging the hot metal into the converter. In external desulphurising pretreatment, a lance is lowered into the molten iron in the ladle and several hundred kilograms of powdered magnesium are added and the sulphur impurities are reduced to magnesium sulphide in a violent exothermic reaction. The sulfide is then raked off. Similar pretreatments are possible for external desiliconisation and external dephosphorisation using mill scale (iron oxide) and lime as fluxes. The decision to pretreat depends on the quality of the hot metal and the required final quality of the steel.

- Filling the furnace with the ingredients is called *charging*. The BOS process is autogenous, i.e. the required thermal energy is produced during the oxidation process. Maintaining the proper *charge balance*, the ratio of hot metal from melt to cold scrap is important. The BOS vessel can be tilted up to 360° and is tilted towards the deslagging side for charging scrap and hot metal. The BOS vessel is charged with steel or iron scrap (25%-30%),if required. Molten iron from the ladle is added as required for the charge balance. A typical chemistry of hotmetal charged into the BOS vessel is: 4% C, 0.2–0.8% Si, 0.08%–0.18% P, and 0.01–0.04% S, all of which can be oxidised by the supplied oxygen except sulphur. (which requires reducing conditions)

- The vessel is then set upright and a water-cooled, copper tipped lance with 3–7 nozzles is lowered into it and high purity oxygen is delivered at supersonic speeds. The lance "blows" 99% pure oxygen over the hot metal, igniting the carbon dissolved in the steel, to form carbon monoxide and carbon dioxide, causing the temperature to rise to about 1700 °C. This melts the scrap, lowers the carbon content of the molten iron and helps remove unwanted chemical elements. It is this use of pure oxygen (instead of air) that improves upon the Bessemer process, as the nitrogen (an undesirable element) and other gases in air do not react with the charge, and decrease efficiency of furnace.

- Fluxes (burnt lime or dolomite) are fed into the vessel to form slag, to maintain basicity above 3 and absorb impurities during the steelmaking process. During "blowing", churning of metal and fluxes in the vessel forms an emulsion, that facilitates the refining process. Near the end of the blowing cycle, which takes about 20 minutes, the temperature is measured and samples are taken. A typical chemistry of the blown metal is 0.3–0.9% C, 0.05–0.1% Mn, 0.001–0.003% Si, 0.01–0.03% S and 0.005–0.03% P.

- The BOS vessel is tilted towards the slagging side and the steel is poured through a tap hole into a steel ladle with basic refractory lining. This process is called *tapping* the steel. The steel is further refined in the ladle furnace, by adding alloying materials to impart special properties required by the customer. Sometimes argon or nitrogen is bubbled into the ladle to make the alloys mix correctly.

- After the steel is poured off from the BOS vessel, the slag is poured into the slag pots through the BOS vessel mouth and dumped.

There are a number of significant improvements, modifications, and process changes of the BOF steelmaking system. For example, when high-phosphorus ore is smelted in the blast furnace, and the BOF is consequently charged with a liquid iron containing more than 0.15 percent of that element, the LD-AC process can be followed, in which lime powder is injected through the lance along with oxygen for quick slag formation. A two-slag practice is then followed for sufficient phosphorus removal, with the first slag runoff being sold for fertilizer. Another variation that finds wide application is the injecting of argon (or sometimes nitrogen) into the molten charge through permeable refractory blocks in the bottom of the converter. Bottom stirring enhances chemical reactions and lowers the steel temperature at the oxygen impact area, resulting in less oxidation of iron and better yield. Another system, called the Q-BOP, uses no top lance at all, blowing oxygen, burnt-lime powder, and, when needed, argon upward through the liquid melt from several gas-cooled or oil-cooled bottom tuyeres. These tuyeres are two concentric steel tubes, with oxygen flowing from the inside annulus and gas or oil flowing through the outer annulus. Cooling of the tubes is accomplished by the endothermic heat required to break down the natural gas or oil into carbon monoxide and hydrogen.

The service life of the bottom of the Q-BOP converter is lower than that of the side wall, thus demanding additional maintenance time for bottom changing. On the other hand, bottom blowing has the advantage of generating a large contact surface among all reactants, thus improving metallurgical reactions and process control. Yield is also higher, since there is less local iron oxidation. However, less oxidation also means the release of less exothermic heat; this decreases the quantity of scrap that can be charged, which can be a cost disadvantage when the price of scrap is low. For this reason, some steel plants enhance bottom blowing with a postcombustion top lance. This is an oxygen lance with additional ports at the tip for burning carbon monoxide into carbon dioxide inside the converter. The additional heat generated by this combined blowing practice increases the potential scrap-charging rate.

Another technology for increasing scrap rates uses an oxy-fuel lance, which preheats the scrap in the converter for about 20 minutes before the liquid blast-furnace iron is added. Another scrap-increasing practice adds aluminum to the charge or melt; this releases heat as it is burned during the oxygen blow. Still another process injects coal powder through a modified oxygen lance or through special bottom tuyeres, simultaneously applying additional oxygen and using a postcombustion lance. In trial operations, this combination has resulted in scrap-charging capabilities all the way up to 100 percent; in other words, no hot metal has been charged, and the converter has become a scrap melter. Increasing scrap-charging rates helps to keep the plant operating when the supply of blast-furnace iron is limited, as, for example, during a blast-furnace reline.

Electric Arc Furnaces

The Electric arc furnace process, or mini-mill, does not involve iron-making. It reuses existing steel, avoiding the need for raw materials and their processing. The furnace is charged with steel scrap, it can also include some direct reduced iron (DRI) or pig iron for chemical balance. The EAF operates on the basis of an electrical charge between two electrodes providing the heat for the process. The power is supplied through the electrodes placed in the furnace, which produce an arc of electricity through the scrap steel (around 35 million watts), which raises the temperature to 1600°C, melting the scrap. Any impurities may be removed through the use of fluxes and draining off slag through the taphole.

Electric arc furnaces do not use coal as a raw material, but many are reliant on the electricity generated by coal-fired power plant elsewhere in the grid. Around 150 kg of coal are used to produce 1 tonne of steel in electric arc furnaces.

The Charge

The major charge material of electric-arc steelmaking is scrap steel, and its availability at low cost and proper quality is essential. The importance of scrap quality becomes apparent when making steels of high ductility, which must have a total maximum content of residuals (i.e., copper, chromium, nickel, molybdenum, and tin) of 0.2 percent. Most of these residuals are present in scrap and, instead of oxidizing during steelmaking, they accumulate and increase in recycled scrap. In such cases some shops augment their scrap charges with direct-reduced iron or cold blast-furnace iron, which do not contain residuals. Generally, the higher contents of carbon, nitrogen, and residuals make the electric-arc process less attractive for producing low-carbon, ductile steels.

Most scrap yards keep various grades of scrap separated. High-alloy shops, such as stainless-steel producers, accumulate, purchase, and charge scrap of similar composition to the steel they make in order to minimize expensive alloying additions.

The Furnace

The electric-arc furnace (EAF) is a squat, cylindrical vessel made of heavy steel plates. It has a dish-shaped refractory hearth and three vertical electrodes that reach down through a dome-shaped, removable roof. The shell diameter of a 10-, 100-, and 300-ton EAF is approximately 2.5, 6, and 9 metres. The shell sits on a hydraulically operated rocker that tilts the furnace forward for tapping and backward for slag removal. The bottom—i.e., the hearth—is lined with tar-bonded magnesite bricks and has on one side a slightly inclined taphole and a spout or, as shown in the figure, an oval hearth and a vertical taphole. With this latter arrangement, a furnace needs be tilted only 10 °For tapping, producing a tight and short tap stream that decreases heat loss and reoxidation of the liquid steel. Before charging, the vertical taphole is closed from the outside by a movable bottom plate and is filled with refractory sand.

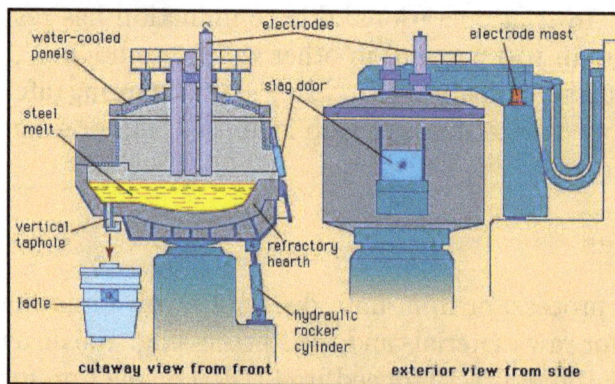

An electric-arc furnace.

Most furnace walls are made of replaceable, water-cooled panels; these are covered inside by sprayed-on refractories and slag for protection and to keep heat loss down. The roof is also made of water-cooled panels and has three circular openings, equally spaced, for insertion of the cylindrical electrodes. Another large roof opening, the so-called fourth hole, is used for off-gas removal. Additional openings in the furnace wall, with water-cooled doors, are used for lance injection, sampling, testing, inspection, and repair. The roof and electrodes can be lifted and moved away for charging scrap and for hearth maintenance.

The graphite electrodes, produced to high standards by a specialized industry, are actually strings of individual electrodes bolted end to end by short graphite nipples. This is done because shorter electrodes are easier to manufacture, transport, and handle. Electrode diameters depend on furnace size; a 100-ton EAF typically uses 600-millimetre electrodes. Three electrode strings are each clamped to arms that extend over the furnace roof and that are bolted to a vertically movable mast located beside the furnace. The mast controls the distance between each electrode tip and the scrap or melt, thereby regulating the arc length and current flow. Power-supply equipment—normally a step-down transformer, vacuum circuit breakers, a tap changer for electrode voltage control, and a furnace transformer—is installed in a concrete vault a short distance from the furnace. Heavy water-cooled cables and the power-carrying arms connect the furnace transformer with the electrodes.

EAF plants are smaller and less expensive to build than integrated steelmaking plants, which, in addition to basic oxygen furnaces, contain blast furnaces, sinter plants, and coke batteries for the making of iron. EAFs are also cost-efficient at low production rates—e.g., 150,000 tons per year—while basic oxygen furnaces and their associated blast furnaces can pay for themselves only if they produce more than 2,000,000 tons of liquid steel per year. Moreover, EAFs can be operated intermittently, while a blast furnace is best operated at very constant rates. The electric power used in EAF operation, however, is high, at 360 to 600 kilowatt-hours per ton of steel, and the installed power system is substantial. A 100-ton EAF often has a 70-megavolt-ampere transformer.

The Process

An arc furnace pouring out steel into a small ladle car. The transformer vault can be seen at the right side of the picture. For scale, note the operator standing on the platform at upper left. This is a 1941-era photograph and so does not have the extensive dust collection system that a modern installation would have, nor is the operator wearing a hard hat or dust mask.

Scrap metal is delivered to a scrap bay, located next to the melt shop. Scrap generally comes in two main grades: shred (whitegoods, cars and other objects made of similar light-gauge steel) and heavy melt (large slabs and beams), along with some direct reduced iron (DRI) or pig iron for chemical balance. Some furnaces melt almost 100% DRI.

The scrap is loaded into large buckets called baskets, with "clamshell" doors for a base. Care is taken to layer the scrap in the basket to ensure good furnace operation; heavy melt is placed on top of a light layer of protective shred, on top of which is placed more shred. These layers should be present in the furnace after charging. After loading, the basket may pass to a scrap pre-heater, which uses hot furnace off-gases to heat the scrap and recover energy, increasing plant efficiency.

The scrap basket is then taken to the melt shop, the roof is swung off the furnace, and the furnace is charged with scrap from the basket. Charging is one of the more dangerous operations for the EAF operators. A lot of potential energy is released by the tonnes of falling metal; any liquid metal in the furnace is often displaced upwards and outwards by the solid scrap, and the grease and dust on the scrap is ignited if the furnace is hot, resulting in a fireball erupting. In some twin-shell furnaces, the scrap is charged into the second shell while the first is being melted down, and pre-heated with off-gas from the active shell. Other operations are continuous charging—pre-heating scrap on a conveyor belt, which then discharges the scrap into the furnace proper, or charging the scrap from a shaft set above the furnace, with off-gases directed through the shaft. Other furnaces can be charged with hot (molten) metal from other operations.

After charging, the roof is swung back over the furnace and meltdown commences. The electrodes are lowered onto the scrap, an arc is struck and the electrodes are then set to bore into the layer of shred at the top of the furnace. Lower voltages are selected for this first part of the operation to protect the roof and walls from excessive heat and damage from the arcs. Once the electrodes have reached the heavy melt at the base of the furnace and the arcs are shielded by the scrap, the voltage can be increased and the electrodes raised slightly, lengthening the arcs and increasing power to the melt. This enables a molten pool to form more rapidly, reducing tap-to-tap times. Oxygen is blown into the scrap, combusting or cutting the steel, and extra chemical heat is provided by wall-mounted oxygen-fuel burners. Both processes accelerate scrap meltdown. Supersonic nozzles enable oxygen jets to penetrate foaming slag and reach the liquid bath.

An important part of steelmaking is the formation of slag, which floats on the surface of the molten steel. Slag usually consists of metal oxides, and acts as a destination for oxidised impurities, as a thermal blanket (stopping excessive heat loss) and helping to reduce erosion of the refractory lining. For a furnace with basic refractories, which includes most carbon steel-producing furnaces, the usual slag formers are calcium oxide (CaO, in the form of burnt lime) and magnesium oxide (MgO, in the form of dolomite and magnesite). These slag formers are either charged with the scrap, or blown into the furnace during meltdown. Another major component of EAF slag is iron oxide from steel combusting with the injected oxygen. Later in the heat, carbon (in the form of coke or coal) is injected into this slag layer, reacting with the iron oxide to form metallic iron and carbon monoxide gas, which then causes the slag to foam, allowing greater thermal efficiency, and better arc stability and electrical efficiency. The slag blanket also covers the arcs, preventing damage to the furnace roof and sidewalls from radiant heat.

Once the scrap has completely melted down and a flat bath is reached, another bucket of scrap can be charged into the furnace and melted down, although EAF development is moving towards single-charge designs. After the second charge is completely melted, refining operations take place to check and correct the steel chemistry and superheat the melt above its freezing temperature in preparation for tapping. More slag formers are introduced and more oxygen is blown into the bath, burning out impurities such as silicon, sulfur, phosphorus, aluminium, manganese, and calcium, and removing their oxides to the slag. Removal of carbon takes place after these elements have burnt out first, as they have a greater affinity for oxygen. Metals that have a poorer affinity for oxygen than iron, such as nickel and copper, cannot be removed through oxidation and must be controlled through scrap chemistry alone, such as introducing the direct reduced iron and pig iron mentioned earlier. A foaming slag is maintained throughout, and often overflows the furnace to pour out of the slag door into the slag pit. Temperature sampling and chemical sampling take place via automatic lances. Oxygen and carbon can be automatically measured via special probes that dip into the steel, but for all other elements, a "chill" sample—a small, solidified sample of the steel—is analysed on an arc-emission spectrometer.

Once the temperature and chemistry are correct, the steel is tapped out into a preheated ladle through tilting the furnace. For plain-carbon steel furnaces, as soon as slag is detected during tapping the furnace is rapidly tilted back towards the deslagging side, minimising slag carryover into the ladle. For some special steel grades, including stainless steel, the slag is poured into the ladle as well, to be treated at the ladle furnace to recover valuable alloying elements. During tapping some alloy additions are introduced into the metal stream, and more lime is added on top of the ladle to begin building a new slag layer. Often, a few tonnes of liquid steel and slag is left in the furnace in order to form a "hot heel", which helps preheat the next charge of scrap and accelerate its meltdown. During and after tapping, the furnace is "turned around": the slag door is cleaned of solidified slag, the visible refractories are inspected and water-cooled components checked for leaks, and electrodes are inspected for damage or lengthened through the addition of new segments; the taphole is filled with sand at the completion of tapping. For a 90-tonne, medium-power furnace, the whole process will usually take about 60–70 minutes from the tapping of one heat to the tapping of the next (the tap-to-tap time).

The furnace is completely emptied of steel and slag on a regular basis so that an inspection of the refractories can be made and larger repairs made if necessary. As the refractories are often made from calcined carbonates, they are extremely susceptible to hydration from water, so any suspected leaks from water-cooled components are treated extremely seriously, beyond the immediate concern of potential steam explosions. Excessive refractory wear can lead to breakouts, where the liquid metal and slag penetrate the refractory and furnace shell and escape into the surrounding areas.

Advantages for Steelmaking

The use of EAFs allows steel to be made from a 100% scrap metal feedstock. This greatly reduces the energy required to make steel when compared with primary steelmaking from ores.

Another benefit is flexibility: while blast furnaces cannot vary their production by much and can remain in operation for years at a time, EAFs can be rapidly started and stopped, allowing the steel mill to vary production according to demand.

Although steelmaking arc furnaces generally use scrap steel as their primary feedstock, if hot metal from a blast furnace or direct-reduced iron is available economically, these can also be used as furnace feed.

As EAFs require large amounts of electrical power, many companies schedule their operations to take advantage of off-peak electricity pricing.

A typical steelmaking arc furnace is the source of steel for a mini-mill, which may make bars or strip product. Mini-mills can be sited relatively near to the markets for steel products, and the transport requirements are less than for an integrated mill, which would commonly be sited near a harbour for access to shipping.

Other Electric Arc Furnaces

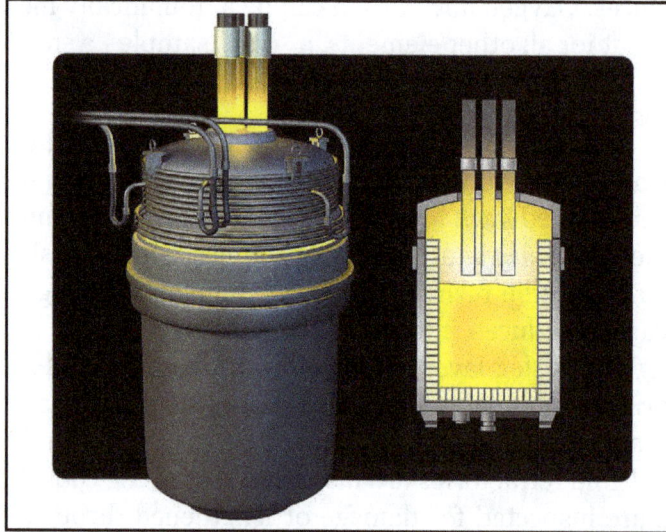

Rendering of a ladle furnace, a variation of the electric arc furnace used for keeping molten steel hot.

For steelmaking, direct current (DC) arc furnaces are used, with a single electrode in the roof and the current return through a conductive bottom lining or conductive pins in the base. The advantage of DC is lower electrode consumption per ton of steel produced, since only one electrode is used, as well as less electrical harmonics and other similar problems. The size of DC arc furnaces is limited by the current carrying capacity of available electrodes, and the maximum allowable voltage. Maintenance of the conductive furnace hearth is a bottleneck in extended operation of a DC arc furnace.

In a steel plant, a ladle furnace (LF) is used to maintain the temperature of liquid steel during processing after tapping from EAF or to change the alloy composition. The ladle is used for the first purpose when there is a delay later in the steelmaking process. The ladle furnace consists of a refractory roof, a heating system, and, when applicable, a provision for injecting argon gas into the bottom of the melt for stirring. Unlike a scrap melting furnace, a ladle furnace does not have a tilting or scrap charging mechanism.

Electric arc furnaces are also used for production of calcium carbide, ferroalloys and other non-ferrous alloys, and for production of phosphorus. Furnaces for these services are physically different from steel-making furnaces and may operate on a continuous, rather than batch, basis. Continuous process furnaces may also use paste-type, Søderberg electrodes to prevent interruptions due to electrode changes. Such a furnace is known as a submerged arc furnace because the electrode tips are buried in the slag/charge, and arcing occurs through the slag, between the matte and the electrode. A steelmaking arc furnace, by comparison, arcs in the open. The key is the electrical resistance, which is what generates the heat required: the resistance in a steel-making furnace is the atmosphere, while in a submerged-arc furnace the slag or charge forms the resistance. The liquid metal formed in either furnace is too conductive to form an effective heat-generating resistance.

Amateurs have constructed a variety of arc furnaces, often based on electric arc welding kits contained by silical blocks or flower pots. Though crude, these simple furnaces can melt a wide range of materials, create calcium carbide, etc.

Cooling Methods

Smaller arc furnaces may be adequately cooled by circulation of air over structural elements of the shell and roof, but larger installations require intensive forced cooling to maintain the structure within safe operating limits. The furnace shell and roof may be cooled either by water circulated through pipes which form a panel, or by water sprayed on the panel elements. Tubular panels may be replaced when they become cracked or reach their thermal stress life cycle. Spray cooling is the most economical and is the highest efficiency cooling method. A spray cooling piece of equipment can be relined almost endlessly; equipment that lasts 20 years is the norm. However while a tubular leak is immediately noticed in an operating furnace due to the pressure loss alarms on the panels, at this time there exists no immediate way of detecting a very small volume spray cooling leak. These typically hide behind slag coverage and can hydrate the refractory in the hearth leading to a break out of molten metal or in the worst case a steam explosion.

Non-pressurized cooling system.

Plasma Arc Furnace

A plasma arc furnace (PAF) uses plasma torches instead of graphite electrodes. Each of these torches consists of a casing provided with a nozzle and an axial tubing for feeding a plasma-forming gas (either nitrogen or argon), and a burnable cylindrical graphite electrode located within the tubing. Such furnaces can be referred to as "PAM" (Plasma Arc Melt) furnaces. They are used extensively in the titanium melt industry and similar specialty metals industries.

Vacuum Arc Remelting

Vacuum arc remelting (VAR) is a secondary remelting process for vacuum refining and manufacturing of ingots with improved chemical and mechanical homogeneity.

In critical military and commercial aerospace applications, material engineers commonly specify VIM-VAR steels. VIM means Vacuum Induction Melted and VAR means Vacuum Arc Remelted. VIM-VAR steels become bearings for jet engines, rotor shafts for military helicopters, flap actuators for fighter jets, gears in jet or helicopter transmissions, mounts or fasteners for jet engines, jet tail hooks and other demanding applications.

Most grades of steel are melted once and are then cast or teemed into a solid form prior to extensive forging or rolling to a metallurgically sound form. In contrast, VIM-VAR steels go through two more highly purifying melts under vacuum. After melting in an electric arc furnace and alloying in an argon oxygen decarburization vessel, steels destined for vacuum remelting are cast into ingot molds. The solidified ingots then head for a vacuum induction melting furnace. This vacuum remelting process rids the steel of inclusions and unwanted gases while optimizing the chemical composition. The VIM operation returns these solid ingots to the molten state in the contaminant-free void of a vacuum. This tightly controlled melt often requires up to 24 hours. Still enveloped by the vacuum, the hot metal flows from the VIM furnace crucible into giant electrode molds. A typical electrode stands about 15 feet (5 m) tall and will be in various diameters. The electrodes solidify under vacuum.

For VIM-VAR steels, the surface of the cooled electrodes must be ground to remove surface irregularities and impurities before the next vacuum remelt. Then the ground electrode is placed in a VAR furnace. In a VAR furnace the steel gradually melts drop-by-drop in the vacuum-sealed chamber. Vacuum arc remelting further removes lingering inclusions to provide superior steel cleanliness and further remove gases such as oxygen, nitrogen and hydrogen. Controlling the rate at which these droplets form and solidify ensures a consistency of chemistry and microstructure throughout the entire VIM-VAR ingot. This in turn makes the steel more resistant to fracture or fatigue. This refinement process is essential to meet the performance characteristics of parts like a helicopter rotor shaft, a flap actuator on a military jet or a bearing in a jet engine.

For some commercial or military applications, steel alloys may go through only one vacuum remelt, namely the VAR. For example, steels for solid rocket cases, landing gears or torsion bars for fighting vehicles typically involve the one vacuum remelt.

Vacuum arc remelting is also used in production of titanium and other metals which are reactive or in which high purity is required.

Secondary Steelmaking

Secondary steelmaking is a critical step in the steel production process between the primary processes and casting. Secondary steelmaking is most commonly performed in ladles.

The Ladle

An open-topped cylindrical container made of heavy steel plates and lined with refractory, the ladle is used for holding and transporting liquid steel. Here all secondary metallurgical work takes place, including deslagging and reslagging, electrical heating, chemical heating or cooling with scrap, powder injection or wire feeding, and stirring with gas or with electromagnetic fields. The ladle receives liquid steel during tapping while sitting on a stand beneath the primary steelmaking furnace. It is moved by cranes, ladle cars, turntables, or turrets. A ladle turret has two liftable forks, usually 180° apart, that revolve around a tower, each fork capable of holding a ladle. Ladles have two heavy trunnions on each side for crane pickup. Support plates under each trunnion are used for setting the ladles onto stands or ladle cars.

The Shell

The side wall of a ladle is slightly cone-shaped, with the larger diameter on top for easy removal of a skull—i.e., solidified steel and slag. A ladle capable of holding 200 tons of steel has an outside diameter of approximately four metres and is about five metres high. Inside the ladle there is usually a 60-millimetre-thick refractory safety lining next to the shell. The working lining, that part contacting the steel and slag, is 180 to 300 millimetres thick, depending on ladle size and location in the ladle. The lining thickness and type of brick in one ladle are often different to counteract increased wear at certain locations—for example, at the impact area of the tapping stream or at the slag line. This results in more equal wear on the ladle lining and an extended ladle service life.

Sometimes, fired clay bricks are used because they bloat—that is, they expand during heating and seal the joints between them. Their thermal shock resistance is high, but their resistance to slag corrosion is low, so that the working lining has to be replaced every 6 to 12 heats. Because ladle rebricking takes about eight hours, up to 12 ladles are sometimes in use in large steelmaking shops in order to assure availability. For ladle operations requiring longer holding times, higher-grade refractory linings are made of high alumina or magnesia bricks. These give greater slag resistance, but they do not bloat and are less resistant to thermal shock. For these reasons, they are kept hot at special preheating stations. Ladles that use these bricks have service lives of up to 80 heats, so that fewer ladles are required. Preheating also decreases the heat loss of liquid steel during tapping and holding.

Tapping

Except for very small ladles, which pour over the lip and a spout or through a teapot arrangement when tilted, most ladles have a funnel-shaped nozzle with a closing device installed in the bottom. Depending on ladle size, these nozzles have an orifice diameter of 15 to 100 millimetres and are made of high-grade refractory material. Often they are opened and closed by a vertical steel stopper rod, which is enclosed in refractory sleeves and partly immersed in the liquid steel. The head of the stopper rod closes the nozzle and is lifted a specific distance for controlling the flow rate; on top it is connected to a vertical slide that is either manually operated by a lever or remotely controlled from the crane pulpit.

Many shops use a slide-gate nozzle, which consists, in principle, of a fixed upper and a movable lower refractory plate. Both plates have holes that are adjusted relative to each other for closed, throttled, and full-open position. The lower plate is hydraulically shifted and is usually replaced after every heat. In a similar system, an old plate is pushed out by a new plate while pouring, and flow control is accomplished by using bottom plates with different orifice diameters. Having the entire flow-control system on the outside of the ladle and the inside of the ladle completely unrestricted is necessary for operating with long holding times and for certain steel treatments conducted in the ladle.

Stirring and Storing

Ladles are often built with one or more permeable refractory bottom blocks and argon hookups for gas stirring. Ladles can also be placed against an electromagnetic stirring coil installed on a ladle car; in this case, their shells are made of a nonmagnetic alloy.

A number of shops use ladle lids to limit the liquid-steel heat loss. Lid-handling systems are normally mechanized, and removing, storing, and placing lids onto the ladles is done automatically.

Ladle Metallurgy

The carrying out of metallurgical reactions in the ladle is a common practice in practically all steel-making shops, because it is cost-efficient to operate the primary furnace as a high-speed melter and to adjust the final chemical composition and temperature of the steel after tapping. Also, certain metallurgical reactions, for reasons of equipment design and operation, are more efficiently performed in the ladle. The simplest form of steel treatment in the ladle takes place when the mixing effect of the tapping stream is used to add deoxidizers, slag formers, and small amounts of alloying agents. These materials are either placed into the ladle before tapping or are injected into the tapping stream.

Controlling Temperature

Deoxidation reactions carried out in the ladle are exothermic and thus raise the temperature of the liquid steel, but the steel also loses heat by radiation from the top surface, by heating of the ladle lining, and by heat flux through the lining and shell. Temperature drops that take place when just holding the steel can range from 0.3° to 2 °C per minute. (Small ladles, owing to their high surface-to-volume ratio, have a greater temperature loss than large ladles.) The rate of temperature drop then slows as the refractories become heated and a steady flow of heat prevails through the lining and slag layer.

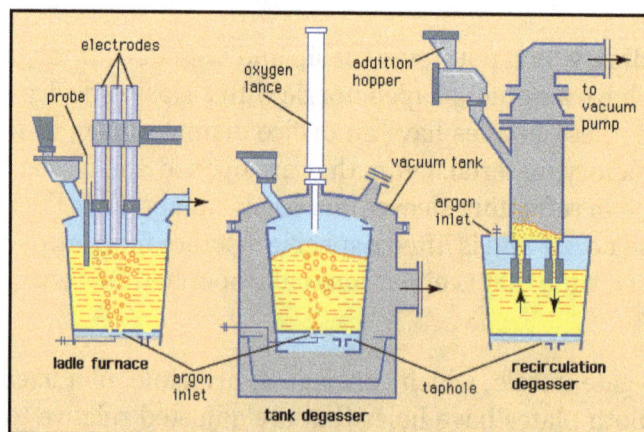

Three ladle treatment stations.

Tapping at the right temperature is necessary in order to meet critical temperature windows for teeming or casting operations. Heat losses during and after tap can usually be predicted by computer, using a process model that considers the temperature and configuration of the tap stream, the thermal condition of the ladle before tap, the thicknesses of the ladle lining and slag layer, the expected holding times and stirring conditions, and the thermal effects of alloying additions. Actual control over steel temperature can be achieved in a ladle furnace (LF). This is a small electric-arc furnace with an 8- to 25-megavolt-ampere transformer, three electrodes for arc heating, and the ladle acting as the furnace shell—as shown in A in the figure. Argon or electromagnetic stirring is applied for better heat transfer. Most LFs can raise the temperature of the steel by 4 °C per minute, and several shops accomplish an increase of 4° to 6 °C by inducing a strong exothermic chemical reaction (for instance, by feeding aluminum and injecting oxygen) at the stirring station. Subsequent argon stirring removes most of the alumina inclusions formed by this process. Both heating technologies permit long holding times of full ladles and improve the continuous caster operation.

Slag Removal

Keeping furnace slags on the molten steel too long can result in a reversion of elements such as phosphorus back into the steel. To avoid this, slag can be removed at slag-skimming stations, where the ladle is tilted forward and a rake scrapes the slag into a slag pot parked beneath the ladle. Some shops use a vacuum system, which sucks the slag off the liquid steel and granulates it instantaneously. In either case, after slag removal the steel is covered with slag formers or an insulating layer to minimize heat loss and reoxidation. Special equipment is used to quickly place a blanket of material on the steel surface.

Stirring and Injecting

In most continuous casting operations, it is necessary to maintain minimal fluctuation in steel temperature, and this requires the use of a ladle stirring station to establish a uniform temperature and chemical composition throughout the ladle. The steel can be stirred by argon injected through a refractory-lined lance or through a permeable refractory block in the bottom of the ladle, or it can be stirred by an electromagnetic coil.

Additions are usually made at the stirring station by a wire feeder, which runs a heavy wire at controlled speed through a refractory-covered lance and into the steel. Aluminum wire is often used for trimming; other materials, such as calcium-silicon, zirconium, and rare-earth metals, are often enclosed in thin steel tubes and are fed by the same machines. The wires and filled tubes are normally shipped to steel plants in large coils, but there are also machines that fill the tubes with the appropriate materials on-site.

Another widely used treatment is powder injection. Powdered metal is fluidized by argon in a pressure vessel and injected by a refractory-lined lance deep into the liquid steel. Because powder has a large contact surface area, it reacts quickly with the steel. Deep injection is beneficial when adding materials such as calcium or magnesium, which evaporate at steelmaking temperature, because ferrostatic pressure suppresses the evaporation of these metals for some time. Powders are shipped to the shop in sealed containers or in special tank cars topped with inert gas.

Desulfurizing

Many powder-injection stations are used for desulfurization. One effective desulfurizer is a calcium-silicon alloy containing 30 percent calcium. Metallic calcium desulfurizes by forming the very stable compound calcium sulfide (CaS), and it is alloyed with silicon because pure calcium reacts instantaneously with water and is therefore difficult to handle. Injecting four kilograms of calcium-silicon per ton of steel can remove approximately three-quarters of the sulfur, so that the sulfur content will drop, for example, from 0.016 to 0.004 percent. For steel grades that do not permit silicon additions, a magnesium-lime mixture is used. Magnesium is a good desulfurizer, and it also acts as a deoxidizer by combining locally with dissolved oxygen. This makes it possible for the lime to desulfurize the steel according to the following reaction:

$$CaO + FeS \rightarrow CaS + FeO.$$

Like magnesium, lime has a double function, because it helps to prevent the very low-melting magnesium powder from melting inside the lance.

Adding calcium accomplishes another important function. Sulfur is normally present in solidified steel in the form of manganese sulfide inclusions, which are soft at hot-rolling temperatures and are rolled into long strings or platelets. This results in poor physical properties of the steel in directions perpendicular to that of the rolling. The addition of calcium improves these properties by forming strong inclusions, containing mainly calcium sulfide, that are not plastic at hot-rolling temperatures. This phenomenon, called inclusion shape control, can also be achieved by small additions of zirconium or rare earth.

Vacuum Treatment

Exposing steel to vacuum conditions has a profound effect on all metallurgical reactions involving gases. First, it lowers the level of gases dissolved in liquid steel. Hydrogen, for example, is readily removed in a vacuum to less than two parts per million. Nitrogen is not as mobile in liquid steel as hydrogen, so that only 15 to 30 percent is typically removed during a 20-minute vacuum treatment.

Another important process is vacuum decarburization and deoxidation. In theory, oxygen and carbon, when dissolved in steel, react to form carbon monoxide until they reach equilibrium at the following relationship:

$$\% C \times \% O = 0.0025 \times CO \text{ pressure}.$$

This means that, under vacuum conditions (when there are only small amounts of carbon monoxide in the surrounding gas and therefore little carbon monoxide pressure), carbon and oxygen will react vigorously until they reach equilibrium at very low levels. For instance, liquid steel at 1 atmosphere pressure may contain 0.043 percent carbon and 0.058 percent oxygen, but, if the pressure is lowered to 0.1 atmosphere, the two elements will react until they reach equilibrium at 0.014 percent carbon and 0.018 percent oxygen. Under a pressure as low as 0.01 atmosphere, equilibrium will be reached at 0.004 percent carbon and 0.006 percent oxygen. In practical operation, the obtainable levels of carbon and oxygen are far above equilibrium conditions, because the movement of carbon and oxygen atoms in liquid steel is time-consuming and treatment time is limited. In addition, the steel is continuously reoxidized by multiple sources of oxygen. Nevertheless, it is common practice to produce ultralow-carbon steel, containing less than 0.003 percent carbon, in 20 minutes at a vacuum treatment station under pressure of one torr. (In vacuum technology, pressures are often expressed in torr, which is equivalent to the pressure of a column of one millimetre of mercury. One atmosphere equals 760 torr.)

There are several types of vacuum treatment, their use depending on steel grade and required production rates. In the tank degasser the ladle is placed in an open-top vacuum tank, which is connected to vacuum pumps. The vacuum pumping system often consists of two or three mechanical pumps, which lower the pressure to about 0.1 atmosphere, and four or five stage steam ejectors, which bring the pressure to under 1 torr, or 0.0013 atmosphere. Practical treatment time is 20 to 30 minutes. The ladles used in tank degassing stations are large and, when filled with steel, retain about one metre of freeboard in order to contain the melt during a vigorous boil.

A modification of the tank degassers is the vacuum oxygen decarburizer (VOD), which has an oxygen lance in the centre of the tank lid to enhance carbon removal under vacuum. The VOD is often used to lower the carbon content of high-alloy steels without also overoxidizing such oxidizable alloying elements as chromium. This is possible because, in the pressure-dependent carbon-oxygen reaction outlined above, oxygen reacts with carbon before it combines with chromium. The VOD is often used in the production of stainless steels.

There are also tank degassers that have electrodes installed like a ladle furnace, thus permitting arc heating under vacuum. This process is called vacuum arc degassing, or VAD.

For higher production rates (e.g., 25 ladles treated per day) and large ladles (e.g., 200 tons), a recirculation degasser is used, as shown in C in the figure. This has two refractory-lined snorkels that are part of a high, cylindrical, refractory-lined vacuum vessel and are immersed in the steel. As the system is evacuated, atmospheric pressure pushes the liquid steel through the snorkels and up into the vessel. One atmosphere lifts liquid steel about 1.3 metres. Injecting argon into one of the snorkels then circulates the steel through the vessel, continuously exposing a portion of the steel to the vacuum. Recirculation facilities are often very elaborate, using fast vessel-exchange systems or even two operating vessels at one station to achieve high production rates. Some units also inject oxygen during vacuum treatment, through either the side or the top of the vessel. This is done to speed up decarburization or, by simultaneously adding aluminum, to increase the steel temperature. Some shops apply a similar system but use a vacuum vessel with only one snorkel. Here, a portion of the steel in the ladle flows in and out of the vacuum vessel and is exposed to the vacuum by a continuous raising and lowering of either the vessel or the ladle.

Argon-oxygen Decarburization

In the production of stainless steel and other high-alloy grades that contain highly oxidizable elements such as chromium, lowering the levels of carbon by regular oxygen injection has the undesirable consequence of oxidizing the alloying elements as well. The argon-oxygen decarburization (AOD) process alleviates this problem by diluting the injected oxygen with argon. This lowers the partial pressure of oxygen and carbon monoxide, so that, based on the pressure-dependent equilibrium relationship $\%C \times \%O = 0.0025 \times CO$ pressure, the oxygen prefers to combine with carbon and oxidizes only a small amount of alloy.

The Converter

The AOD process is carried out in a refractory-lined converter similar to the BOF but with two to six argon-oxygen tuyeres installed in the lower side wall. The tuyeres consist of two concentric steel tubes, with the inert gas flowing in the outer annulus and oxygen in the inner tube. The converter has tilting and emission-control equipment similar to that of the BOF; the lining is also basic, but it lasts only 50 to 100 heats because of the long refining time and the high temperature of more than 1,700 °C (3,100 °F) that is necessary for improving the chromium yield. Most shops have three converter shells and one trunnion ring at a blowing station, rotating them between operation, relining, and preheating.

The Process

When making austenitic stainless steel, the AOD converter is charged with liquid high-carbon chromium-nickel steel that has been melted in a regular EAF and may contain 1.5 percent carbon, 19 percent chromium, and 10 percent nickel. The blow starts with a high-oxygen gas mixture of, for instance, 80 percent oxygen and 20 percent argon, because there is still plenty of carbon in the steel with which oxygen prefers to combine. As the carbon level drops, the gas mixture is gradually changed into one rich in argon; this may end with a blowing gas of 20 percent oxygen and 80 percent argon. After a blowing time of about one hour, the final carbon content is on the order of 0.015 percent, and only about 2 percent chromium has been lost. The steel is then deoxidized by ferrochrome silicon and desulfurized with burnt lime. Argon is also blown during this end phase for better mixing and removal of hydrogen and nitrogen.

The tap-to-tap time is about two hours, and consumption of oxygen and argon is about 25 and 20 cubic metres, respectively, per ton of steel. To minimize cost, argon is sometimes replaced by nitrogen or compressed air at the beginning of the blow. AOD converters with capacities up to 160 tons are in operation.

Casting of Steel

Ingot Pouring

The simplest way to solidify liquid steel is to pour it into heavy, thick-walled iron ingot molds, which stand on stout iron plates called stools.

Solidification Processes

During and after pouring, the walls and bottom of the mold extract heat from the melt, and a solid shell forms, growing approximately with the square root of time multiplied by a constant. The value of the constant depends on the heat flux between the already solidified shell and the cooling media surrounding it and is actually equivalent to the solidified shell's thickness after one minute—namely, about 20 millimetres when solidifying steel. Accordingly, the ingot shell is about 40 millimetres thick after four minutes and 60 millimetres after nine minutes. As the shell thickens, the level of the liquid melt in the centre of the mold drops, because solidified steel has a higher density than liquid steel—i.e., 7.86 versus 7.06 grams per cubic centimetre (4.5 versus 4.1 ounces per cubic inch). This creates a cavity on top of the ingot, as shown in A in the figure by a schematic presentation of solidifying layers. Since an open cavity oxidizes, it does not weld during hot rolling and must be cut off, resulting in a loss of steel. The cavity can be made shallower by keeping the top of the ingot hot and liquid longer. This is done by inserting insulating refractory heads and by adding exothermic powders; more liquid steel can also be added after a good-sized shell has formed.

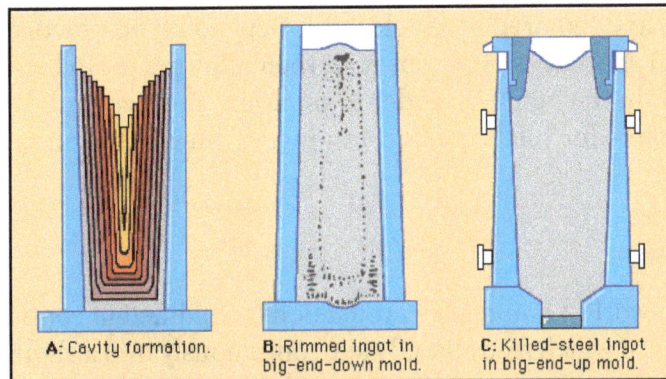

A: Cavity formation.   B: Rimmed ingot in big-end-down mold.   C: Killed-steel ingot in big-end-up mold.

Ingot solidification

The solidification pattern described above can be observed in well-deoxidized steel, which shows no evolution of gas as it solidifies. For this reason, it is called a killed steel. A different solidification pattern is applied to certain other steels to which fewer deoxidizers have been added. These contain a controlled amount of dissolved oxygen, which, during solidification, reacts with carbon and generates a mild carbon monoxide boil. The rising carbon monoxide bubbles stir the melt, lift inclusions, and cause the formation of a very clean shell about 50 millimetres thick, called the rim. After the rim has formed, a cooling plate is placed on top of the ingot, freezing a layer of liquid steel and trapping the gas bubbles inside the solidifying ingot, as shown in B in the figure. This ingot has no open cavity, but there are many blowholes in the centre that normally weld together during hot-rolling. Low-carbon steel, because of its higher dissolved oxygen content, is often cast this way and is called rimmed steel. Normally, rimmed steel is cast into a big-end-down mold, as shown in B in the figure, for easier mold stripping and ingot handling.

An important characteristic of all solidification processes is segregation. This takes place when crystals grow in a multicomponent melt, because crystals are always purer than the liquid melt from which they solidify. Therefore, as steel solidifies, the levels of carbon, phosphorus, and sulfur grow in the remaining liquid, resulting in an enrichment of these elements in the centre of the ingot. Segregation can be minimized by keeping segregating elements at low levels or by solidifying at a fast rate—i.e., by not providing the time for separation. It is also impaired by stirring the melt.

Pouring Procedures

The layouts of pouring pits differ greatly, depending on the type of steel produced and the rate of production. In top pouring conducted in high-tonnage shops, a row of perhaps 20 molds is lined up in buggies on a railroad track in front of a pouring platform. A crane brings the ladle to the platform and holds it while the operator fills one mold after another. After standing for a specified time, the molds are pulled out of the teeming aisle and into a stripper building, where they are lifted from the ingots. In a different procedure, called bottom pouring, as many as six ingot molds stand on a single large and thick bottom plate with several pipelike refractory runners installed on its top surface. These runners connect the molds to a refractory-lined, funnel-shaped feeder tube, which receives liquid steel from the ladle and directs it to the molds, filling them simultaneously from the bottom. Bottom pouring avoids the splashing from the ladle stream that is experienced during top pouring. The system is often completely mechanized, with the bottom plates movable on wide transfer tracks and prepared for the next use away from the pouring aisle.

Iron molds are cleaned and repaired in a mold yard. Depending on practice, they are replaced after 40 to 70 pours. Most specialty steel shops pour their alloy grades in big-end-up molds and use hot tops, as shown in C in the figure, in order to minimize the size of the cavity and consequent steel loss. All large ingots—for instance, 200-ton ingots intended for forgings—are also poured this way.

Continuous Casting

About 55 percent of the world's liquid steel production is solidified in continuous casting processes, the most widely used of which feeds liquid steel continuously into a short, water-cooled vertical copper mold and, at the same time, continuously withdraws the frozen shell, including the liquid steel it contains.

Tundish, Mold and Secondary Zone

The key control parameter of continuous casting is matching the flow of liquid steel into the mold with the withdrawal speed of the strand out of the mold. The control of flow rates is accomplished by the tundish, a small, refractory-lined distributer that is placed over the mold and that receives steel from the furnace ladle. Withdrawal speed is controlled by driven rolls, which contact the strand at a point where it has already developed a thick, solidified shell.

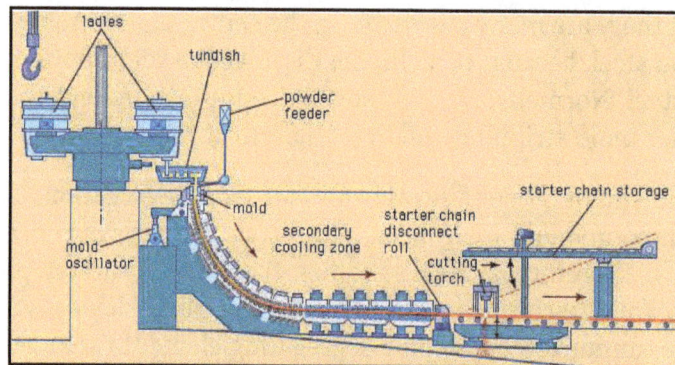

A curved-mold continuous slab caster.

Feeding of the caster mold from the tundish is controlled by a stopper rod or a sliding gate similar to the equipment used in ladles. The liquid steel in the tundish must be within a specific temperature "window"—a range just above its liquidus that is determined by the steel's grade; in addition, measures are always taken to keep air away from the steel in order to minimize reoxidation. Shielding can be accomplished by pouring steel through refractory tubes that are immersed in the steel or through wide sleeves that are pressurized with argon. The tundish itself is covered with a lid and is often also topped with argon. Both ladle and tundish sit on a turret or transfer car to permit a quick exchange.

The mold is made of copper because of the high heat conductivity of that metal. It is heavily water-cooled and oscillates up and down to avoid sticking of the solidified shell to its walls. In addition, the mold wall is lubricated by oil or slag, which is maintained on the steel meniscus and flows down into the gap between mold and strand. The slag layer, when used, is formed by the continuous addition of casting powder. Besides providing lubrication, it keeps air away from the liquid steel, acts as a heat barrier, and absorbs inclusions.

Many continuous casters contain sensors in the mold for automatically synchronizing the flow of liquid steel into the mold with the strand withdrawal speed. As it exits the mold, the strand has a shell thickness of only about 10 millimetres and is immediately water-cooled by spray nozzles. The strength and soundness of the shell at this location determine the maximum casting speed, because rupturing it would result in a breakout of liquid steel and damage to the caster. On its way down, the strand is supported by many rolls to avoid a bulging of the shell by the ferrostatic pressure of the liquid steel it contains. As the shell thickness increases toward the end of this so-called secondary cooling zone, the supporting rolls grow larger and are spaced farther apart. The secondary zone is often also called the metallurgical length, because this is where the strand solidifies and the cast structure develops. Depending on the strand's cross section and the casting speed, it can be 10 to 40 metres long. The flow of water to the many nozzles in the various sections is often computer-controlled and automatically adjusted as casting conditions change.

After the strand passes through the last pair of support rolls, it enters the run-out table and is cut, while moving, by one or two oxyacetylene torches.

Design Principles

Continuous casters in commercial operation are built according to different design principles. For some steels and solidification patterns, all components are arranged in a vertical line—a straight mold, a straight secondary cooling zone, and vertical strand cutting. Other casters also have a straight mold and a vertical secondary cooling zone, but they bend the strand on its way down, after it has solidified, into a horizontal direction and cut it on a run-out table. (In spite of the horizontal turn, even this design requires a high building and a long ladle lift.)

The majority of continuous casters have a curved mold, a curved secondary cooling zone, and a series of straightening rolls before the horizontal run-out table. Everything down to the straightener is on one radius or on several matching radii. This design results in a low casting machine, as shown in the figure.

Billet, Bloom, Beam and Slab

Different design principles are used for casting strands of different cross sections. Billet casters solidify 80- to 175-millimetre squares or rounds, bloom casters solidify sections of 300 by 400 millimetres, and beam blank casters produce large, dog-bone-like sections that are directly fed into an I-beam or H-beam rolling mill. Huge slab casters solidify sections up to 250 millimetres thick and 2,600 millimetres wide at production rates of up to three million tons per year.

In order to match the quantity of steel produced in a heat with the solidification capacity of a mold for a certain strand section, it is often necessary to use a multistrand caster. Some billet casters have six molds in one line next to one another, and all are fed from the same tundish.

Casting Procedures

To begin casting, a starter head matching the inside dimension of the mold and connected to a starter chain is moved up into the mold. The starter chain has dimensions similar to the strand to be cast and is long enough to be moved up and down by the driven rolls. When liquid steel fills the mold, it freezes to the caster head, which is immediately withdrawn. The chain in front of the solidifying strand moves through the secondary cooling zone, and, after the head has cleared the last support roll, it is disconnected from the strand by an upward-moving push-out roll. The chain is then pulled by a winch onto a support cradle, lifted from the table, and stored for reuse. At the end of casting, when the tundish is almost empty, the flow of steel to the mold is discontinued, and the strand is stopped and, after solidifying, completely withdrawn. For the next cast, the starter chain, with the head in front, is moved again by the driven rolls into the secondary cooling zone and mold.

Casting of one ladle takes 45 to 90 minutes, depending on heat size, steel grade, caster layout, and casting conditions. Turning the caster around—that is, preparing it for the next cast—is usually accomplished in a half hour, but it takes longer when the mold is changed for casting a different section. Slab casters often use molds with movable side plates, thus permitting a fast change of width during caster turnaround or even during casting. Such devices, together with fast exchange systems for casting tubes, tundishes, and ladles, permit sequential heats to be cast without stopping the caster—sometimes for several days. Starting and stopping a caster causes a few metres of steel on both ends of the strand to fall below the specified properties, thereby lowering the steel-to-strand yield. In sequential casting, on the other hand, the yield from liquid steel to acceptable strand approaches 100 percent, compared with perhaps 93 percent when turning the caster around after each ladle or to 86 percent in an ingot-casting operation that uses a blooming or slabbing mill to roll a slab or bloom of the same size. The benefits are substantial because much less raw material, liquid steel, and energy are needed to make the same tonnage of cast product.

Metallurgical quality is often enhanced by computer control over some or all systems of the caster. Casting conditions are often further improved by electrical tundish heating to adjust steel temperature, by electromagnetic stirring coils around the strand to decrease segregation, by in-line rolling to compact the centre just before it solidifies, and, most important, by well-designed inspection systems to check the liquid steel and the hot strand during casting. Such systems provide a high level of quality assurance, making it possible to charge the cut strand hot into a reheat furnace or, with only a little reheating of the edges, directly into a hot-rolling mill. This not only minimizes reheating but eliminates cooling, cold inspection, scarfing or grinding, and storage. Plants that integrate a continuous caster with a hot-rolling mill often need only 90 minutes to convert liquid steel into a hot-rolled product.

Variations

Some plants have been built specifically for direct rolling. One example is a thin-slab caster that casts strands 50 millimetres thick and 1,250 millimetres wide at speeds of about five metres per minute. After the strand is cut on the run-out table, the slabs are directly heated in-line in a long tunnel furnace or by induction coils and then fed, also in-line, directly into the finishing train of a hot-strip mill. With everything in one continuous line, operating and maintenance systems must be kept at the highest level.

Another special continuous process is the rotary casting of rounds, mainly for seamless tubes. A rotary caster is similar to a straight-mold vertical caster, except that the round mold, the strand, and the withdrawal system revolve at about 75 rotations per minute. This creates a centrifugal force within the strand and results in a cleaner cast and better contact between strand and mold. Still another variation is the casting of rounds in a horizontal caster. This entirely different system employs a large tundish with a horizontal nozzle in its side wall that extends directly into a water-cooled horizontal mold. The strand oscillates and is pulled out of the mold in small increments each time a new shell has formed at the mold entrance. Everything is located on one level, so that there are no high ladle lifts. Ferrostatic pressure in the strand is also very low, but segregation tendencies caused by gravitational forces require more careful preparation of the liquid steel.

There have been, and still are, many continuous-casting concepts tested in laboratories, pilot plants, and trial operations. Examples include single- or dual-roll strip casters, which cast strip directly from liquid steel, and belt casters for thin-slab production. There have also been hundreds of patents issued on continuous casting, all with the goal of making the process more cost-efficient, improving metallurgical control, and casting as close to the final product shape as possible.

Special Solidification Processes

For the manufacture of special products, refining and solidification processes are often combined.

Vacuum Ingot Pouring

Vacuum ingot pouring is often employed to produce very large ingots that are subsequently processed, in expensive forging and machining operations, into such products as rotors for power generators. In this process, an ingot mold is placed inside a cylindrical tank that is connected to vacuum pumps. The tank is closed by a lid, and a small, stopper-operated ladle having a capacity of about 25 tons of liquid steel is set on top of the lid. The nozzle of this so-called pony ladle is sealed by an aluminum disk, the tank is evacuated, and the furnace ladle starts pouring steel into the pony ladle. When the ferrostatic pressure reaches a certain point, the stopper is opened, the aluminum plate burns through, and the stream of liquid steel is degassed before it fills the mold for solidification. Pouring under vacuum lowers the hydrogen content, an important matter for large ingots.

Vacuum Arc Remelting (VAR)

In this process, employed for casting steels that contain easily oxidized alloying elements, a consumable electrode made of forged steel or of compacted powder or sponge is continuously melted by an arc under vacuum. At the same time, the shallow molten pool underneath the electrode is continuously solidified in a water-cooled, normally round copper mold. As the mold is filled, the electrode moves up. The melting current, in flowing between the electrode and the mold, passes through the arc, liquid pool, and solidified strand. Melting under high vacuum lowers the levels of dissolved oxygen, oxide inclusions, hydrogen, nitrogen, and elements having a high vapour pressure, such as lead, manganese, and tin. In addition, the shallow pool results in a directional solidification, with the crystals growing parallel to the axes of the ingot; this greatly improves the subsequent hot-forming operation. There is no segregation and no cavity. Ingots weighing up to 50 tons and measuring 1.5 metres in diameter have been cast with this method.

Electroslag Remelting (ESR)

In this process, there is a slowly melting consumable electrode and a water-cooled mold for solidification, as in vacuum arc remelting, but the melting is conducted under normal atmosphere and is accomplished by a thick, superheated layer of slag on top of the shallow metal pool. This slag is resistance-heated by the high electrical current passing from the electrode to the mold, and it also desulfurizes the molten steel drops as they pass through on their way from the electrode to the liquid pool. Solidification patterns are similar to those in vacuum arc remelting. The ingot surface is very clean, owing to the presence of a slag layer between the ingot and mold, and does not need surface conditioning. Some electroslag installations cast ingots heavier than 200 tons.

Steel Foundry

Foundries that cast steel into commercial products mainly employ coreless induction furnaces or electric-arc furnaces for melting scrap. Scrap quality is normally high because a large portion of return scrap is used in the form of gates and risers left over from previous casting operations. Since it is often not necessary to refine scrap—that is, to lower the sulfur and phosphorus content—an acid process can be applied using a high-silica slag that may contain 60 percent silica, 10 percent lime, 10 percent manganese oxide, and 15 percent iron oxide. This permits the furnaces to run with a cheaper acid lining.

Tapping temperatures are usually higher than for ingot pouring or continuous casting in order to have a liquid steel with good fluidity that fills the thin parts of a casting. Molding is similar to that in gray-iron foundries, but a more heat-resistant mold material is necessary because of the higher temperatures. Solidifying steel castings normally show a higher linear shrinkage (1.5 percent) than gray iron castings, which shrink about 1 percent. Small parts are cast in greensand molds, but larger parts are made in stronger dry-sand molds.

Benefits of Continuous Casting

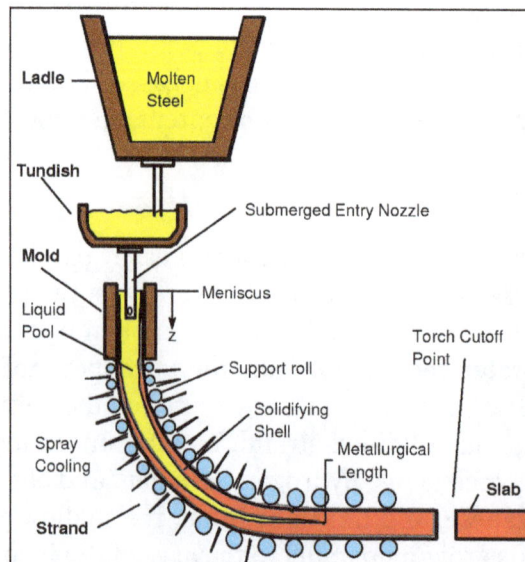

Unlike other processes of casting, the time line of steps in continuous casting is entirely different. While in other casting processes, each step of casting heating of the metal, poring of the molten liquid into casts, solidification and cast removal are a sequential process, in continuous casting all steps occur congruently and hence it saves a lot of processing time.

Continuous casting is a method that was invented to enhance the production of metals. The continuity of the casting helps to lower the cost of the casted steel. Further, it helps in the standardized production of steel cast. Further, the carefully controlled process also reduces errors leading to better quality steel casts. They increase the productivity and produce better yields.

Continuous casting eliminates some of the problems of traditional casting methods. For example, they eliminate piping, structural and chemical variations that are common problems of ingot casting method. All the casting products manufactured by continuous casting possess uniform properties. The solidification rate of the molten metal is also ten times faster than the solidification of the metal in ingot casting method.

Continuous casting has several advantages but it is also a process that needs distinct resources. This is the reason why this process is employed only in industries that require high yield of steel cast.

Forming and Treating of Steel

Principles

Forming processes convert solidified steel into products useful for the fabricating and construction industries. The objectives are to obtain a desired shape, to improve cast steel's physical properties (which are not suitable for most applications), and to produce a surface suitable for a specific use. During plastic forming, the large crystals in cast steel are converted into many small, long crystals, transforming the usually brittle cast into a ductile and tough steel. In order to accomplish this, it is often necessary to reduce the cross section of a cast structure to one-eighth or even less of its original.

The major forming processes are carried out hot, at about 1,200 °C (2,200 °F), because of steel's low resistance to plastic deformation at this temperature. This requires the use of reheating furnaces of different designs. Cold forming is often applied as a secondary process for making special steel products such as sheet or wire.

There are a number of steel-forming processes—including forging, pressing, piercing, drawing, and extruding—but by far the most important one is rolling. In this process, the rolls, working always in pairs, are driven in opposite directions with the same peripheral velocity and are held at a specific distance from each other by heavy bearings and mill housings. The steel workpiece is pulled by friction into the roll gap, which is smaller than the cross section of the workpiece, so that both rolls exert a pressure and continuously form the piece until it leaves the roll gap with a smaller section and increased length. As shown in the figure, the reduction in cross section is calculated by subtracting the out-section (S_2) from the in-section (S_1) and then dividing by S_1. Assuming the workpiece maintains its original volume as it is formed, the elongation (L_2) divided by the original length (L_1) equals S_1 divided by S_2. When rolling flat products, there is not much change in width, so that the thickness alone can be used to calculate reduction.

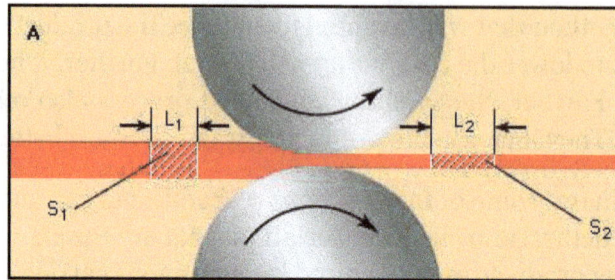

Gap between two rolls, showing reduction and elongation of workpiece.

The basic principles of a rolling-mill design are shown in B in the figure. Two heavy bearings mounted on each side of a roll sit in chocks, which slide in a mill housing for adjusting the roll gap with a screw. The two housings are connected to each other and to the foundation, and the complete assembly is called a roll stand. There are also compact rolling units, which do not have housings; often used in the tandem rolling of long products, they can be exchanged quickly for repair or for a change in the rolling program. Rolls are driven through spindles and couplings, either directly or via a gear, by one or several electric motors. Depending on the product rolled, there are stands that have two, three, four, and more rolls; accordingly, they are given the names two-high, three-high, four-high, six-high, cluster mill, and planetary mill. For rolling strip, heavy backup rolls support the smaller work rolls, because thin rolls form flat material better than do large-diameter rolls.

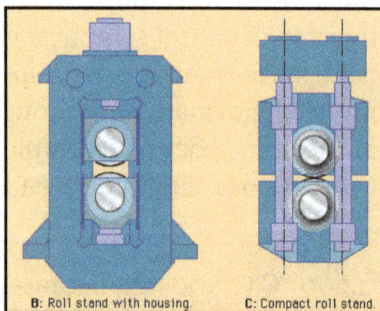

Two basic rolling-mill designs.

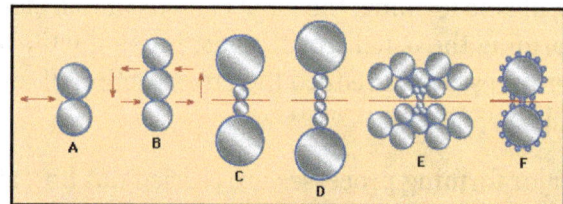

Two-high, three-high, four-high, six-high, cluster, and planetary roll arrangements.

In a rolling shop, stands are arranged according to three layout principles. One is called the open train in which the stands are arranged side by side, often driven by the same motor and linked by spindles. This arrangement is applied only to the rolling of long products, with guides or cross-transfers being used to move the workpiece from stand to stand. A tandem mill arrangement has one stand behind the other and is used for high-production rolling of almost all products. This continuous arrangement requires the construction of long rolling trains and buildings, but layouts can be shortened by a so-called semicontinuous mill, in which the workpiece is passed back and forth through a reversing mill before being sent through the rest of the line. When open-train and tandem arrangements are combined for rolling long products in more compact layouts, it is called a cross-country mill.

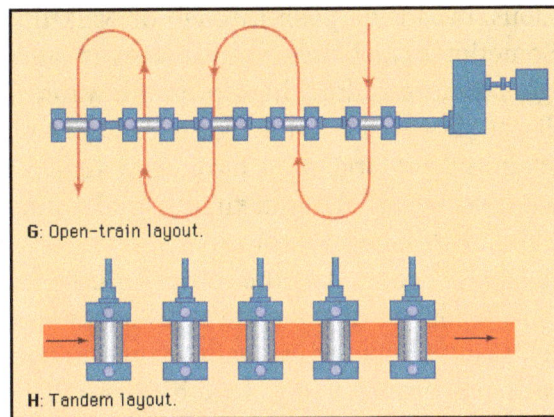

G: Open-train layout.

H: Tandem layout.

Two rolling-mill arrangements.

Slabs and Blooms

Cast ingots, sometimes still hot, arrive at slabbing and blooming mills on railroad cars and are charged upright by a special crane into under-floor soaking pits. These are gas-fired rectangular chambers, about 5 metres deep, in which four to eight ingots are simultaneously heated to about 1,250 °C (2,300 °F). An ingot used for conversion into a slab can be 1.5 metres wide, 0.8 metre thick, and 2.5 metres high and can weigh 23 tons. The soaking pits are highly computerized for scheduling, firing rates, heating times (which can last 8 to 18 hours), and rolling programs.

After heating, a tiltable transfer buggy brings a hot ingot to a two-high reversing mill, which takes one pass after another, reversing the rolls and roller table each time the ingot has passed through. Because each pass reduces the slab by only about 50 millimetres, it may take 21 passes, including several edge passes with the slab standing upright on its edges, to obtain a slab measuring 0.2 metre thick, 1.5 metres wide, and 10 metres long.

The rolls usually have a diameter of about 1.2 metres; each is driven by one or two electric motors totaling 7,000 to 12,000 horsepower. The two roller tables, situated in front and in back of the stand, have movable manipulators that guide the slab into the rolls and turn it onto its edges when required. High-pressure water nozzles remove surface scale, and a crop-shear discards the ends and cuts the slab into proper length. Some slabbing mills place a pair of heavy vertical rolls next to the horizontal rolls for edge rolling; this avoids the time-consuming turning of the slab into an upright position. Such an arrangement is called a universal mill.

For making long products, blooms some 250 millimetres square are rolled from ingots in a similar fashion on the same type of mill.

Plates

Rolled from heavy slabs supplied by a slabbing mill or continuous caster or sometimes rolled directly from an ingot, plates vary greatly in dimensions. The largest mills can roll plates 200 millimetres thick, 5 metres wide, and 35 metres long. These three dimensions are determined by the slab or ingot weight as well as the rolling-mill size. Sometimes only a few plates of the same dimensions and quality specifications are ordered.

Most mills have two continuous, broadside push-through or walk-through furnaces, which heat the slabs to about 1,250 °C. Sometimes two batch-type furnaces are also used for heating odd-sized or extra-heavy slabs and ingots. Before rolling, high-pressure water jets descale the slabs. Most plate mills are four-high mills, as shown in C in the figure, and are supplemented by vertical edge rolls. The work rolls and backup rolls of large mills have diameters of 1.2 and 2.4 metres, respectively, and a roll face length up to 6 metres. Their maximum total rolling force is often 10,000 tons, and their rolls are driven by an 8,000-kilowatt motor. Most mills have hydraulic roll adjustment, which transmits the roll pressure to a computer; the computer uses this and other rolling parameters, such as temperature and thickness of the plate at all locations, to control the rolling process by a mathematical model. This technology—actually a computerized art—permits not only the rolling of huge workpieces with high accuracy (e.g., to a thickness tolerance of 0.2 millimetre) but also the control of rectangularity, flatness, plan-view shape, yield, physical properties, and profile. Several plants are even capable of rolling plates with a tapered or stepped thickness. Sometimes plants use two rolling mills, a roughing stand and a finishing stand, to improve surface quality and increase production. Most plate mills also have elaborate equipment for leveling, cooling, shearing or milling of edges, heat-treating, and marking.

Hot Strip

The rolling of hot strip begins with a slab, which is inspected and, if necessary, surface cleaned either manually or by scarfing machines with oxyacetylene torches. The slabs are then pushed, or walked on their broadside, through gas-fired furnaces that have a hearth dimension of about 13 metres by 30 metres. In a pusher-type furnace, the slabs slide on water-cooled skids, and, each time a new slab is charged, a heated slab drops through a discharge door onto a roller table. In walking-beam furnaces, several walking beams lift the workpieces from the hearth, move them forward, and set them back down in a series of rectangular movements. These furnaces have the advantage of producing no cold stripes and skid marks across the slabs. Preheating temperature, as with slabs and plates, is about 1,250 °C.

A heated slab moves first through a scale breaker, which is a two-high rolling mill with vertical rolls that loosens the furnace scale and removes it with high-pressure water jets. Then the slab passes through four-high roughing stands, typically four arranged in tandem, which roll it to a thickness of about 30 millimetres. The stands are spaced about 30 to 70 metres apart, so that the slab is only in one roll gap at a time. After roughing, it proceeds to a long (about 140 metres) roller table in front of the finishing train for cooling, when required for metallurgical reasons. As the slab enters the finishing train (at about 20 metres per minute), a crop-shear cuts the head and tail, and high-pressure steam jets remove the secondary scale formed during rolling. Six or seven four-high finishing stands then roll the strip to its final thickness of 1.5 to 10 millimetres.

Finishing stands are arranged in tandem, only five to six metres apart and close-coupled, so that the strip is in all rolls at the same time. For process control, a computer receives continuous information from on-line sensors, measuring such parameters as thickness, temperature, tension, width, speed, and shape of the strip, as well as roll pressure, torque, and electrical load. Reduction is high in the first stands (e.g., 45 percent) and low in the last stand (e.g., 10 percent) to ensure good surface and flatness of the strip, which leaves the last finishing stand at 600 to 1,200 metres per minute and 820° to 950 °C (1,510° to 1,750 °F). The strip is water-cooled on a 150-metre-long run-out table and coiled at high speed at 520° to 720 °C (970° to 1,325 °F). Mills have at least two coilers to ensure 100 percent availability.

All the equipment in a hot-strip mill is arranged in a straight line of about 600 metres from furnace to coiler, with the slab or strip passing only once through each stand. Total installed power of only the heavy rolling-mill motors can exceed 125,000 horsepower.

Controlling rolling and coiling temperatures is essential for metallurgical reasons, because it greatly influences the physical properties of both hot-rolled and cold-rolled strip. Also, a number of systems are in use to improve dimensional control of the strip. In order to guide the strip through the flat rolls of a tandem mill, it is made thicker in the centre (by about 0.1 millimetre) than at the edges. This so-called crown, as well as the strip's entire profile, is often controlled by roll bending, accomplished by hydraulic cylinders and extra-long bearings on each side of the extended roll neck. Another system, which improves the wear pattern and service time of the work rolls, is roll shifting—i.e., a sideward adjustment of the rolls along their axes. Normally, the rolling program of a hot-strip mill is influenced by roll wear. Since the heaviest roll wear takes place at the colder edges of the strip, it is common to roll wide strips first and narrow strips later. Roll shifting permits so-called schedule-free rolling—i.e., strip of any width can be rolled at any time. It also is used for controlling the strip profile.

Many highly mechanized hot-strip mills have a capacity of three million to five million tons per year, and as much as 60 percent of the raw steel produced in industrial countries is rolled on these mills. There are, however, hot-strip mills designed for smaller production. For example, a semi-continuous hot-strip mill has only one reversing rougher in front of the finishing train. Another rolling system goes even farther and uses one four-high reversing rougher and one four-high reversing finishing mill, with hot-coiling boxes in front and in back of the finishing mill. (Hot coilers operate in a furnace to keep the strip hot.) In addition, there are planetary-type hot-strip mills, which have a cage of approximately 20 small rolls around each of two backup rolls. The small rolls, in turning around the big roll, make a small reduction every time they pass over the wedge-shaped portion of the workpiece in the roll gap. Planetary mills can reduce a slab from 25 to 2.5 millimetres in one pass—although at a slow rate.

Cold Strip

The rolling of cold strip begins with the retrieval of hot-rolled strip from a coil storage yard, which often uses fully automated cranes for setting and retrieving coils according to rolling schedules. The coils are first descaled in continuous pickle lines,. The cleaned and oiled coils are fed into a cold-reduction mill, which is usually a tandem mill of four to six four-high stands with an uncoiling reel at the entry and a recoiling reel at the exit. When rolling from, for example, 2 millimetres to 0.3 millimetre, the cold reduction is usually 35 percent on the first stands and 15 percent on the final stand. The exit speed is normally high, often 100 kilometres (60 miles) per hour, in order to achieve proper production rates with such small cross sections. Since the strip temperature may go as high as 200 °C (390 °F), proper cooling of strip and rolls is essential. Heavy-duty lubricants are also used to minimize friction in the roll gap.

Typically, the work rolls have a diameter of a half-metre, and the backup rolls of 1.2 metres. For wide strip, the roll face can be 2.4 metres long. The work rolls are precision ground with a specific crown to compensate for roll bending. The last stand usually takes only a small reduction to improve control over the final thickness, profile, and flatness of the strip. To improve control further, many shops use hydraulic roll bending, or they use a differential cooling of the rolls to change their shape by thermal expansion. For additional shape control, a number of shops employ a six-high mill as the last stand, shifting the work rolls and intermediate rolls along their axes during rolling. This provides continuous shape control, because the rolls are ground to a specific profile. All these systems, together with the high speed of rolling, make cold-reduction mills highly complex to operate and controllable only by computer.

Usually, cold-rolled strip cannot be used as rolled, because it is too hard and has low ductility. Therefore, it is annealed in batch or continuous annealing plants. After annealing, the strip is cold-rolled to about a 3-percent reduction on a temper mill to improve its physical properties. (Temper mills are dry, four-high reversing mills that are similar to cold-reduction mills but less powerful.) This rolling operation also gives the strips their final surface finish, an important characteristic and often specified by the customer. If required, shearing lines cut the coils into sheets.

Several plants integrate some or all of the operating steps of a cold-rolling shop into a continuous operation, moving an endless strip (welded together at the pickler or cold mill) through the processes without coiling and coil storage. Indeed, some plants move one continuous strip from the pickle line to the temper-mill exit, with cold-rolling and annealing in between. One of these continuous lines can take less than two hours to convert a hot-rolled coil into a shippable cold-rolled product—a great operating advantage that requires, however, excellent computer control at all levels and perfect maintenance to provide the needed reliability for the completely linked-up equipment. With direct charging of a hot-strip mill from a continuous caster, it is possible to have liquid steel in shippable form five hours after it has been tapped at the furnace.

Billets, Bars and Rods

Billets

Billets are the feedstock for long products of small cross section. In cases when they are not directly cast by a continuous caster, they are rolled from blooms by billet mills. One method of rolling billets, which are usually 75 to 125 millimetres square, is to use a three-high mill with box passes, as shown in A in the figure. After a rectangular bloom is rolled into a square cross section at the lower rolls, it is lifted to the next pass on the upper rolls and rolled back into a rectangular one; this is turned 90° while being lowered on a roller table for another square rolling in the lower pass, and so on. In another method, alternating horizontal and vertical stands are arranged in tandem, using diamond and square passes without turning or twisting the billet.

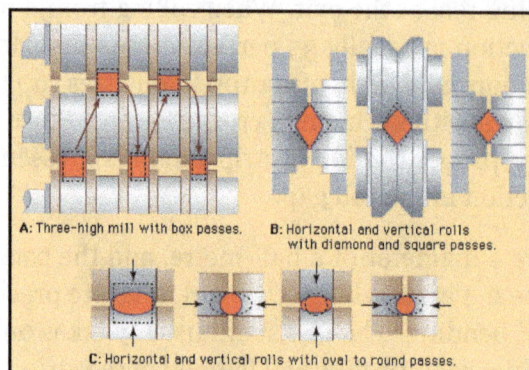

The rolling of billets and bars.

Bars

Bars are long products, usually of round, square, rectangular, or hexagonal cross section and of 12- to 50-millimetre diameter or equivalent. (Since bar mills are also capable of rolling small shaped products such as angles, flats, channels, fence posts, and tees, these products are sometimes called merchant bars.) In rolling bars, a billet measuring, for instance, 120 millimetres square and five metres long is heated in pusher or walking-beam furnaces to 1,200 °C. There is a great variety of layouts used in bar-rolling mills. In principle, after removal of the furnace scale by water jets, a primary reduction takes place in several passes through roll stands in open, semicontinuous, or fully continuous arrangement. These can use an alternating square-diamond rolling principle on horizontal and vertical rolls, as shown in B in the figure, or a series of oval-to-round passes, as illustrated in C in the figure.

Guiding the strands properly from roll gap to roll gap is an important part of this rolling technology. When using only horizontal rolls, the guides also twist the bar 90° between diamond and square passes. In a continuous arrangement of close-coupled mills—in which several roll pairs or roll sets are installed a short distance from one another and all are driven through gears by one or two motors—bars are allowed to buckle in a controlled vertical loop in order to maintain a low tension in a bar between the stands. When using an open-train arrangement, a U-shaped trough called the repeater guides and threads the strand, as indicated in G in the figure. This generates a horizontal loop, caused by the entry speed of each receiving stand being slower than the exit speed of the delivering stands.

The finishing stand of a bar mill gives the bar its final shape and often a specific surface pattern, such as the protrusions on concrete-reinforcing bars. The rolling speed increases as the cross section at each successive stand decreases, and the exit speed can be as high as 15 metres per second. The hot bar is then cut by a flying shear into cooling-bed length (e.g., 50 metres), after which it is cooled, inspected, and cold-cut to shipping length.

Rods

Rod mills are similar to bar mills at the front end, but the finishing end is different. Rods have a smaller section (5.5 to 15 millimetres in diameter) and are always coiled, while bars are normally shipped in cut length. The final rolling in rod mills often takes place in a close-coupled set of 10 pairs of small rolls (200 and 150 millimetres in diameter); these are all installed in a block, with their axes at a 45° angle and arranged in an alternating fashion like the vertical and horizontal rolls in a continuous bar mill. Exit speed of small-diameter rods can go up to 100 metres per second. The rod is immediately coiled by quickly rotating laying heads and cooled before bundling. For enhanced production, two strands are often rolled simultaneously. Such high-speed operation requires cooling of the rod and almost every rolling-mill component. The cooling condition of the bars and rods is also carefully controlled to meet metallurgical specifications.

Computers are used for designing roll passes and for scheduling and controlling the complex operations. Bar and rod mills produce 150,000 to 750,000 tons per year. The largest mills are housed in buildings up to 600 metres long. The most space-consuming part of these manufacturing facilities (and the source of most bottlenecks) is the finishing and shipping area, which handles the many different lightweight shapes that are produced in various steel grades, heat treatments, and surface conditions and are made to many specific customer orders.

Shapes

These are long products with irregular cross sections, such as beams, channels, angles, and rails. Rolling starts with blooms that may be 150 millimetres by 200 millimetres by 5 metres long. The blooms are received, either cold or hot, directly from the blooming mill or continuous caster. They are charged into a pusher or walking-beam continuous furnace and heated for up to three hours to 1,200 °C. (Sometimes, three batch-type furnaces are used instead.)

Most shapes are formed by grooved rolls with mating projections that form together a window in their gap. This window becomes progressively smaller and more like the desired shape, pass after pass, until at the end, in the final pass, the specified cross section is obtained. D in the figure shows only 5 progressive passes out of about 11 in the rolling of a rail. Rolling shapes usually takes a total of 9 to 15 passes, with an area reduction of about 25 percent at the initial passes and only 7 percent at the last pass.

Roll and pass design is critical for this rolling technology. There are usually three to five stands arranged in various ways, each taking one to five passes. Only one pass is made through the finishing stand, which controls the final dimension and surface. Sometimes two-high reversing mills are used at the beginning in a fashion similar to blooming mills, with manipulators on run-out roller tables. In other cases, two or three three-high, nonreversing stands are arranged as an open train; in this arrangement, lifting roller tables move the workpiece between the upper and lower pass lines, and the workpiece is in only one roll gap at a time. Mills that produce medium and small shapes often have stands in tandem arrangement, rolling one workpiece simultaneously in several stands and using a controlled loop between stands. Wide-flange I-beams and H-pilings are usually rolled on universal mills using vertical edgers, as indicated in E in the figure. Blooms with a dog-bone cross section are often supplied to these structural-shape mills by beam-blank continuous casters.

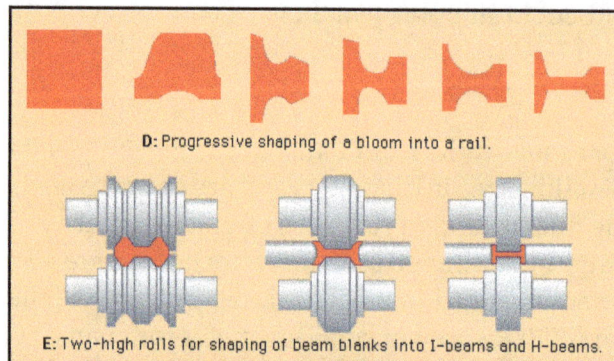

D: Progressive shaping of a bloom into a rail.

E: Two-high rolls for shaping of beam blanks into I-beams and H-beams.

The rolling of structural shapes.

Rolling temperatures are carefully controlled for metallurgical reasons. Heavy-walled, wide-flange I-beams are sometimes heat-treated in-line by computer-controlled water quenching and by tempering with their own retained heat. The heads of rails are often heat-treated in-line to improve wear and impact resistance. Rails are also slow-cooled under an insulated cover, directly after rolling, for at least 10 hours to diffuse hydrogen out of the steel.

After rolling, a hot saw cuts the shapes into lengths that can be handled by the cooling bed. Each shop conducts large-size finishing operations such as straightening, cold-cutting to ordered length, marking, and inspection.

Tubes

Tubular products are manufactured according to two basic technologies. One is the welding of tubes from strip, and the other is the production of seamless tube from rounds or blooms.

Welded Tubes

The most widely used welding system, the electric-resistance welding (ERW) line, starts with a descaled hot-rolled strip that is first slit into coils of a specific width to fit a desired tube diameter. In the entry section is an uncoiler, a welder that joins the ends of coils for continuous operation, and a looping pit, which permits constant welding rates of, typically, three metres per minute. Several consecutive forming rolls then shape the strip into a tube with a longitudinal seam on top, as shown schematically in A in the figure. Two squeeze rolls press the seam together, while two electrode rolls or sliding contacts feed the electric power to the seam for resistance heating and welding. A cutting tool removes the flash created during welding, and, after a preliminary inspection, the tube is cut into cooling-bed length by a saw that moves with the tube.

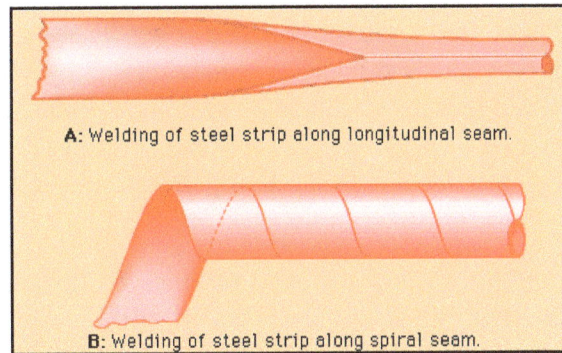

Production of welded tubes.

Tubes up to 500 millimetres in diameter with walls 10 millimetres thick are produced on ERW lines. Larger-diameter pipes are often produced by forming the strip into an endless spiral, as shown schematically in B in the figure. Forming is followed by continuous welding of the seam, often by automatic arc welding. Pipes up to 1.5 metres in diameter and with a 12-millimetre wall thickness are sometimes produced by this spiral welding process. Still larger pipes are produced from plates by a U-ing and O-ing process, which applies heavy presses to form plates into a U and then an O. The longitudinal seam (or seams) are then welded by automatic arc-welding equipment.

Seamless Tubes

Seamless tube rolling always begins by piercing a round or bloom to generate a hollow. In roll piercing, an oval round is preheated to about 1,200 °C and is cross-rolled slowly between two short, large-diameter rolls that rotate in the same direction. The round also revolves and is pulled into the roll gap in a spiraling motion, because the rolls have a converging-diverging shape and are installed relative to each other at an angle of about 20°. This revolving, continuous plastic working of an oval cross section between the two rolls creates tensile stresses in the long axes of the oval, which rupture the centre and create a cavity. At this point the cavity meets the piercer, which is a projectile-shaped rotating cone held in place by a bar and a thrust bearing. The piercer acts like a third roll in the centre and produces the inside of the tube.

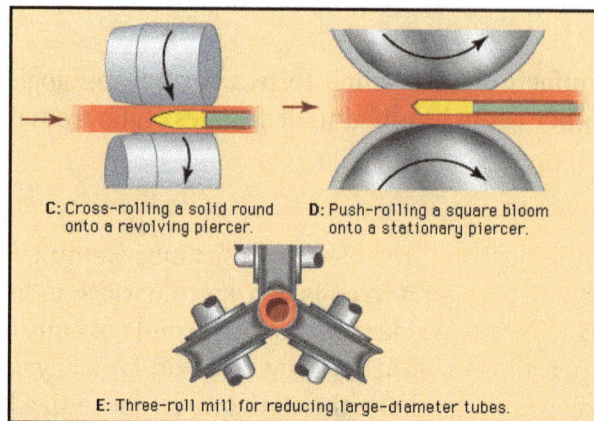

Production of seamless tubes.

The cross or helical rolling action of roll piercing demands excellent hot formability of the pre-rolled round. Another process, push piercing, does not have such exacting requirements. This usually takes continuously cast square blooms and forms them into hollow rounds by the action of a heavy hydraulic pusher, which pushes them into the gap of two large-diameter contoured rolls that form together a circular pass line. In the roll gap the bloom is met by a heavy piercer, which forms the hollow, as shown in D in the figure. This mill can form a 250-millimetre-square, 3-metre-long bloom into a tube with an outside diameter of 300 millimetres and an inside diameter of 150 millimetres. Since there are only compression forces acting on the steel in this process, the workpiece is practically not elongated at all.

A number of rolling technologies are used to form the pierced hollows into tubes with specific dimensions and tolerances. Often, the hollow is reheated and then sent through another cross-roll piercer mill, called the elongator; this reduces the wall thickness by 30 to 60 percent. In a subsequent step, a long, preheated, lubricated cylinder called a mandrel may be inserted into the tube. The tube would then be rolled, with the mandrel inside, in a continuous close-coupled, seven-stand, two-high mill, usually with the rolls arranged at a 45° angle and in an alternating pattern like the horizontal and vertical rolls. A very uniform wall thickness can be formed by this process. Smaller diameter tubes are often formed from larger tubes in a continuous three-roll, close-coupled stretch-reduction mill. These mills sometimes have 20 sets of rolls arranged in tandem.

Open-die Forging

Heavy ingots, some weighing up to 300 tons, are sometimes formed at steel plants by huge hydraulic presses with a forging force of up to 10,000 tons. These make such large products as rotors for power-generating units or large sleeves for rolls or pressure vessels. Careful, uniform heating of the ingots to forging temperature may take 60 hours, and, before completion of the forging process, the workpiece may be reheated six times. The forging is accomplished by flat-, vee-, or swage-shaped dies, depending on the shape of the final product. Saddles and mandrels are used for forging rings and sleeves. The workpiece is connected to a long bar, which helps to move and turn it by a crane or manipulator. Large heat-treating furnaces are available in these forging shops to improve microstructure and to release internal stresses caused by the forging operation.

Wire

The cold drawing of wire is an important and special sector of steelmaking. It produces wire in hundreds of sizes and shapes and within a spectrum of physical properties unmatched by other steel products. Wire is also produced with many types of surface finish.

Treating of Steel

Heat-treating

In principle, heat-treating already takes place when steel is hot-rolled at a particular temperature and cooled afterward at a certain rate, but there are also many heat-treating process facilities specifically designed to produce particular microstructures and properties. The simplest heat-treating process is normalizing. This consists of holding steel for a short time at a temperature 20° to 40 °C above the G-S-K line and then cooling it afterward in still air. Holding the steel in the gamma zone transforms the as-rolled or as-cast microstructure into austenite, which dissolves carbides. Then, during cooling, a very uniform grain is formed, consisting of either pearlite and ferrite or pearlite and cementite, depending on carbon content.

In all heat-treatment operations, the temperatures, holding times, and heating and cooling rates are varied according to the chemical composition, size, and shape of the steel. In general, alloy steels, which have a lower heat conductivity than carbon steels, are heated more slowly to avoid internal stresses.

Annealing

To make steel ductile for subsequent forming operations, an annealing treatment is applied. In annealing, the steel is usually held for several hours at several degrees below Ar1 and then slowly cooled. This precipitates and coagulates the carbides and results in large ferrite crystals. Cold-formed steel is usually annealed and recrystallized in this manner, holding it for several hours at about 680 °C (1,260 °F).

Annealing is performed in an inert or reducing atmosphere to prevent any oxidation of the steel surface. In batch annealing of cold-rolled strip, for example, several coils are set on a base and on top of one another. Then they are covered with a shell made of heat-resistant steel, which is sealed on the bottom and holds the inert gas during annealing. A gas-fired bell furnace is then lowered by a crane over this cover for heating. The total processing time, including cooling, may be 50 to 120 hours, depending on furnace load and steel grade.

In a different system, the cold-rolled strip is pulled through an 80-metre-high furnace with the strip moving up and down between many top and bottom rolls. These continuous-annealing furnaces are usually heated by gas-fired radiation tubes in order to separate combustion gases from the inert atmosphere surrounding the strip. In this dynamic annealing process, the strip is heated to higher temperatures (for example, 780 °C, or 1,440 °F), held for only a few seconds, and immediately cooled by fast-circulating inert gas. The entry and exit sections of continuous-annealing lines are built, as on other strip-processing lines, to allow an uninterrupted and constant travel (at, say, 500 metres per minute) of the strip through the process section—in this case, the heating and cooling zones. The entry group has two uncoiling reels, a cross-shear, welding equipment for joining two strips, and a strip accumulator. The latter is often a looping tower, which supplies the process section above with strip at constant speed while welding is done at the entry section. The exit group works in a similar fashion, with a looping tower and two reels; it also cuts samples and substandard portions out of the strip.

Continuous-annealing lines are often 200 metres long, and the strip between uncoiler and recoiler is more than one kilometre in length. Strip annealed this way is not as soft as batch-annealed steel—a disadvantage compensated for by using ultralow-carbon steels—but it does have operating advantages in that annealing of one coil may take only one hour and the mechanical and surface properties of the strip are very uniform.

Quenching and Tempering

The most common heat treatment for plates, tubular products, and rails is the quench-and-temper process. Large plates are heated in roller-type or walking-beam furnaces, quenched in special chambers, and then tempered in a separate low-temperature furnace. Uniform heating and quenching is crucial; otherwise, residual stresses will distort and warp the plate. Tubes made for very demanding services, such as oil drilling, are usually heat-treated in walking-beam furnaces and special quench-and-temper systems.

The heads of rails are sometimes heat-treated in-line by induction heating coils, air quenching, and tempering by a controlled use of the heat retained in the rail after quenching. Heavy-walled structural shapes are sometimes water-quenched directly after the last pass at the rolling mill and also tempered by the heat retained in the steel. In-line heat-treating results in cost savings because it eliminates extra heat-treating processes and facilities.

The quenching media and the type of agitation during quenching are carefully selected to obtain specified physical properties with minimum internal stresses and distortions. Oil is the mildest medium, and salt brine has the strongest quenching effect; water is between the two. In special cases, steel is cooled and held for some time in a molten salt bath, which is kept at a temperature either just above or just below the temperature where martensite begins to form. These two heat treatments are called martempering and austempering, and both result in even less distortion of the metal.

Surface Treating

The surface treatment of steel also begins during hot-rolling, because reheating conditions, in-line scale removal, rolling temperature, and cooling rate all determine the type and thickness of scale formed on the product, and this affects atmospheric corrosion, paintability, and subsequent scale-removal operations. Sometimes the final pass in hot-rolling generates specific surface patterns—for example, the protrusions on reinforcing bars or floor plates—and in cold-rolling a specific surface roughness is rolled into the strip at the temper mill to improve the deep-drawing operation and to assure a good surface finish on the final product—for instance, on the roof of an automobile.

Pickling

Before cold forming, hot-rolled steel is always descaled, most commonly in an operation known as pickling. Scale consists of thin layers of iron oxide crystals, of which the chemical compositions, structures, and densities vary according to the temperature, oxidizing conditions, and steel properties that are present during their formation. These crystals can be dissolved by acids; normally, hot hydrochloric or sulfuric acid is used, but for some alloy steels a different acid, such as nitric acid, is needed. In addition, inhibitors are added to the acid to protect the steel from being dissolved as well.

The pickling of hot-rolled strip is carried out in continuous pickle lines, which are sometimes 300 metres long. The strip is pulled through three to five consecutive pickling tanks, each one 25 to 30 metres long, at a constant speed of about 300 metres per minute. Like other continuous strip-processing lines, pickle lines also have an entry and exit group to establish constant pickling conditions. After the last acid tank, there are sections that rinse, neutralize, dry, inspect, and oil the strip.

Long products, such as bars and wire rods, are normally pickled in batch operations by placing them on racks and immersing them in long, acid-containing vats. Sometimes shotblasting is used instead of pickling; this removes scale from heavy hot-rolled products by directing high-velocity abrasives onto the surface of the steel.

Cleaning

The removal of organic substances and other residues from the surface of steel, in particular after cold forming with lubricants, is carried out either in special cleaning lines or in the cleaning sections of another processing line. Hot solutions of caustic soda, phosphates, or alkaline silicates are used. The strip is often moved through several sets of electrodes, which, submerged in the cleaning liquid, electrolytically generate hydrogen gas at the steel surface for lifting residues off the strip.

Surface Coating

Approximately one-third of the steel shipped by the industry is coated on its surface by a metallic, inorganic, or organic coating. By far the largest installations are operated for coating cold-rolled strip. In this group the most widely used are those which coat the steel with zinc, zinc alloys, or aluminum.

In hot-dip galvanizing lines, which also have the usual entry and exit groups, the strip moves first at constant speed—say, 150 metres per minute—through a cleaning section and a long, horizontal, nonoxidizing preheating furnace. (When hard strips are coated directly after cold reduction, this furnace is also used for annealing.) The hot strip, still protected by the inert furnace atmosphere in a long steel channel, enters the zinc bath at a temperature of approximately 480 °C (900 °F), supplying heat to the zinc bath, which is at about 440 °C (825 °F). The liquid zinc is contained in a refractory-lined, induction-heated vessel called the zinc pot. When it contacts the strip surface, the liquid zinc alloys with the iron and forms a strong metallurgical bond. However, the iron-zinc alloy is brittle, so that the coating, if too thick, will crack during forming of the sheet. For this reason, about 0.1 to 0.25 percent aluminum is added to the zinc, inhibiting iron-zinc formation and keeping the alloy layer to less than 15 percent of the total coating thickness. Excess liquid zinc is wiped off each side of the strip by two gas-knives, which have long, slotlike orifices through which high-pressure gas is blown. Coating thickness is controlled by adjusting the gas pressure and the location of the knives. Common coating weights are 180 or 275 grams of zinc per square metre of sheet, counting both surfaces. Sometimes, a heavy coating is produced on one side and a lighter coating on the other; this is called a differential coating. The total length of hot-dip galvanizing lines, including furnaces and cooling zones, sometimes reaches 400 metres. The entire system is computer-controlled, based on the continuous, in-line measuring of the coating weight.

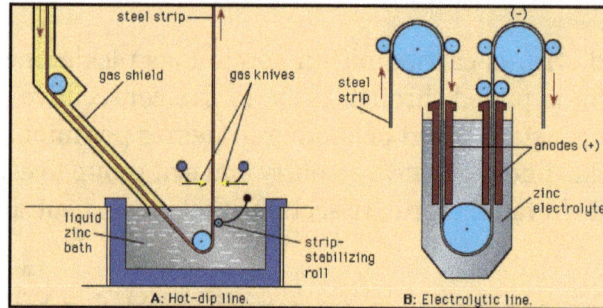

Principles of (A) hot-dip and (B) electrolytic galvanizing.

There are several variations of the basic galvanizing process. The galvanneal process heats the strip above the zinc pot right after coating, using induction coils or gas-fired burners to create a controlled, heavy iron-zinc layer for improved weldability, abrasion resistance, and paintability of the product. Several processes use a zinc-aluminum alloy, and some lines have a second pot filled with liquid aluminum for aluminum coating. The pots are often quickly exchangeable.

Electrolytic galvanizing lines have similar entry and exit sections, but they deposit zinc in as many as 20 consecutive electrolytic coating cells. Of the several successful cell designs, The strip, connected to the negative side of a direct current through large-diameter conductor rolls located above and between two cells, is dipped into a tank of electrolyte by a submerged sink roll. Partially submerged anodes, opposing the strip, are connected to the positive side of the electric current by heavy bus bars. Zinc cations (i.e., positively charged zinc atoms) present in the electrolyte are converted by the current into regular zinc atoms, which deposit on the strip. The bath is supplied with zinc cations either by zinc anodes, which are continuously dissolved by the direct current, or by zinc compounds continuously added to the electrolyte. In the latter case the anodes are made of insoluble materials, such as titanium coated with iridium oxide. The electrolyte is an acidic solution of zinc sulfide or zinc chloride with other bath additions to improve the quality of the coating and the current efficiency. Coating thickness is easier to control here than in the hot-dip process because of the good relationship between electrical current and deposited zinc. Theoretically, 1.22 kilograms of zinc are formed when applying a current of 1,000 amperes over one hour; this means that a line with an installed electrical capacity of one million amperes can deposit 1.22 tons of zinc per hour. The control parameters of such a line are mainly the current density between anodes and strip, the line voltage, the chemical composition and temperature of the electrolyte, and the line speed.

Electrolytic lines normally produce lower coating weights (15 to 60 grams per square metre) than do hot-dip lines, and they can also easily supply differential coatings and one-sided coatings for specific applications. Many lines can deposit zinc-alloy coatings, such as zinc-nickel or zinc-iron, and some lines are capable of producing multilayered coatings of different alloys, the goal being to optimize a combination of specific requirements such as corrosion resistance, weldability, abrasion resistance, drawability, and paintability. The processing speed of electrolytic galvanizing lines can often reach 180 metres per minute.

Electrolytic tinning lines for the production of tinplate are, in principle, of similar design, except that all rolls are smaller (because the strip is thinner and narrower), the line speed is faster (e.g., 700 metres per minute), and different electrolytes and anodes are used. Electrolytic coating lines also coat strips with chromium and other metals and alloys. Most of these lines have a shear line installed at the end to produce cut-to-length sheets upon request.

Many long products are also surface coated. Wires, for example, are often hot-dip galvanized in continuous multistrand lines. In addition, electrolytic coating of wire with all types of metal is often done by hanging coils from current-carrying C-hooks or bars into long vats, which have anodes installed and are filled with electrolyte. Many tubular products and reinforcing bars are coated with organic material to inhibit corrosion.

References

- Iron-processing, technology: britannica.com, Retrieved 19 June, 2019

- Primary-steelmaking, steel, technology: britannica.com, Retrieved 12 April, 2019

- How-steel-produced, uses-coal, coal: worldcoal.org, Retrieved 10 August, 2019

- Basic-oxygen-steelmaking, steel, technology: britannica.com, Retrieved 20 June, 2019

- Basic-oxygen-steelmaking, steel, technology: britannica.com, Retrieved 29 August, 2019

- Electric-arc-steelmaking, steel, technology: britannica.com, Retrieved 27 February, 2019

- H. W. Beaty (ed.), Standard Handbook for Electrical Engineers, 11th Ed., McGraw Hill, New York 1978, ISBN 0-07-020974-X pages 21.171-21.176

- Secondary-steelmaking, steel, technology: britannica.com, Retrieved 7 May, 2019

- Casting-of-steel, steel, technology: britannica.com, Retrieved 11 July, 2019

- Basics-of-continuous-casting-of-steel: calmet.com, Retrieved 6 January, 2019

- Forming-of-steel, steel, technology: britannica.com, Retrieved 18 March, 2019

Applications of Steel

Steel is a versatile material which is used in diverse areas. Some of them are construction, automotive and transport sectors, packaging food and catering, and energy production. The chapter closely examines the applications of steel in these areas to provide an extensive understanding of the subject.

Applications of Steel in Construction

Steel is a very important character in construction work. It can apply in concrete, farming work, beam, column, etc. Steel strong in tensile and compressive, so when apply in column, it can help column to easily crack or damage. Compare with wood make building, steel make building allow people run away from the building when it was in fire.

Steel can be totally recycled, and it will not produce any waste when it apply in construction. Talk about the advantage, steel is strong and it can last longer than any material. Carbon fiber has same properties as steel, but cost of carbon fiber is extremely expensive compare with steel.

There is no perfect material in the world, even steel is good, but it can corrode anytime when we expose to air or water. Client or owner spent a lot of money for the further treatment. So rather to spend money on the further treatment, they choose to do something on it before it install. Such way like coating, material resist, and catholic protection.

The main uses of steel in construction are:

1. Steel frames: Steel frames are the high-end structural components of modern housing and building design. These very strong frames are able to manage huge loads and cover large areas cost-effectively. Modern CAD design using universal beams allows beams to be designed precisely, including everything from loads and stresses to dimensions and positioning. Steel frames are an ongoing revolution in construction, constantly pushing the limits and allowing for bigger and taller buildings and better quality of construction.

2. Prefabricated steel: Modern steel prefabrication can deliver a wide range of construction solutions that are custom made to your design specifications. Precision cutting, welding and high-quality steel are a winning combination in these demanding roles. Whether you're building a trailer, house or even a shopping centre, prefabrication is the new, cost-efficient approach to traditional construction.

3. Roofing: The humble metal roof now offers a whole range of new advantages. Better steel manufacturing, the protection of Zincalume combined with modern colours and better cost-efficiencies have kept steel roofing as a preferred option. In the domestic market, steel roofing is now an architectural fashion statement, offering a chance to redesign your home

and add an upper level on a good cost base. Modern steel roofing can deliver excellent value in terms of space, product life, looks, reliability and even energy saving.

4. Outbuildings: Steel outbuildings are creating a new species of outbuildings that are much more elegant and affordable than their predecessors. Universal beams and universal columns make building a lot easier than with timber frames. These high-quality buildings are springing up all over Australia, partly because of better costs but mainly because of tremendous flexibility in design.

5. Custom building: In Australia's very competitive construction industry, cost and custom design options are drivers for demand. Steel is the best option in many custom design scenarios. A steel design can be literally anything. If you've got the beams and columns, you can design a palace and cost it to the last cent. The flexible range of modern steel designs means fabulous choices for modern homes, extensions, renovations and much more.

Applications of Steel in Transport

Steel provides strong, safe and sustainable transport solutions. Steel facilitates our mobility and the transport of goods. Whether in the form of bicycles, motorcycles, cars, buses, trains, ships or planes – or in the transport networks that support them – steel is essential to every mode of transport. Continuously reinforced concrete roadways are structurally supported by steel rebar and help to improve fuel efficiency for large vehicles.

Steel is well-suited to transport applications because it is durable, strong (providing safety in the case of collision), lightweight, UV-resistant, affordable, and 100% recyclable. Innovations in design and the development of new high-strength steels have also played a key role in improving the efficiency of many of these transport modes whilst at the same time considerably reducing life cycle greenhouse gas (GHG) emissions.

How Steel is used in Transport

Including automotive, around 16% of steel produced worldwide is used to meet society's transport needs. Steel is also essential to the related infrastructure: roads, bridges, ports, stations, airports and fueling. Some major applications include:

For Ships and Shipping Containers

Shipbuilding traditionally uses structural steel plate to fabricate ship hulls. Modern steel plates have much higher tensile strengths than their predecessors, making them much better suited to the efficient construction of large container ships. A particular type of plate is available with a designed resistance to corrosion, ideal for building oil tankers. Such steels make possible much lighter vessels than before, or larger-capacity vessels of the same weight, offering significant opportunities to save on fuel consumption and hence CO_2.

Steel ships transport 90% of the world's cargo. An estimated number of 17 million containers of varying types made up the worldwide container fleet and the majority are made of steel.

For Trains and Rail Cars

Rail transport requires steel in the trains and for the rails and infrastructure. For short or medium haul journeys, rail reduces travel times and CO_2 emissions per passenger kilometer compared to nearly all other forms of transport.

Steel makes up 15% of the mass of high speed trains and is essential.The main steel components of these trains are bogies (the structure underneath the trains including wheels, axels, bearings and motors). Freight or goods wagons are made almost entirely of steel.

For Aeroplanes

Steel is required for the landing gear.

For Infrastructure

Transport networks: steel is used in bridges, tunnels, rail track, and in constructing buildings such as fueling stations, train stations, ports and airports. About 60% of steel use in infrastructure is rebar. The rest is sections and rail track.

Stainless Steels Application in Automotive and Transport Sectors

Automotive and Transport sectors are making increasing use of stainless steels to reduce weight, improve aesthetics, enhance safety and minimize life cycle cost. Characterized by superior fire and corrosion resistance, they ensure safety and reliability. Since stainless steels exhibit superior combination of high strength, ductility, formability and toughness compared to other metals and alloys, the intrinsic weight of vehicle decreases and its load carrying capacity and fuel efficiency increases. Maintenance cost is naturally lower and stainless steel component at the end of its long life is easily recycled.

In automobiles, stainless steels are most extensively used for exhaust systems. To improve efficiency, the designs for these components are becoming very complex and performance criteria are getting tougher. Since corrosion resistance remains vital for this application, appropriate titanium, niobium or dual stabilized grades with very low interstitial content are used with choice of grade depending on operating conditions.

Buses, trams make extensive use of austenitic stainless steels for outer panels. On account of transformation induced plasticity, these steels develop very high strength coupled with good formability in cold rolled tempers and constitute ideal material for structural components. The exceptional strength to weight ratio and energy absorption capacity enables the designers to reduce weight and enhance crash worthiness while ensuring longer life span due to superior corrosion resistance.

As new austenitic, ferritic and duplex stainless steels are evolving, automotive and transport industry is intensively exploring their potential.

Automotive Advanced Steels

One of the important tasks for the 21st century is the maintaining of sound ecology. The reduction of the burden on the environment is an inevitable task assigned to industry. A direct contribution of the steel industry to the reduction of fuel consumption is the supply of steels enabling the lightening of automotive weight. There is a direct correlation between kerb weight of vehicle and fuel consumption. As a thumb rule, 10% of weight reduction will lead to 3%-7% less of fuel consumption. To reduce the weight of the vehicle and thus energy consumption, the usage of high strength steel in auto-making is gradually increasing. Figure shows the diagram representing the materials used for manufacturing cars. The trend showing that there is increase in kerb weight in the year 2000, this is due to implementation of stringent safety laws. Following that there is decrease in kerb weight. This weight reduction was possible by development of HSS and AHSS and stringent environment law.

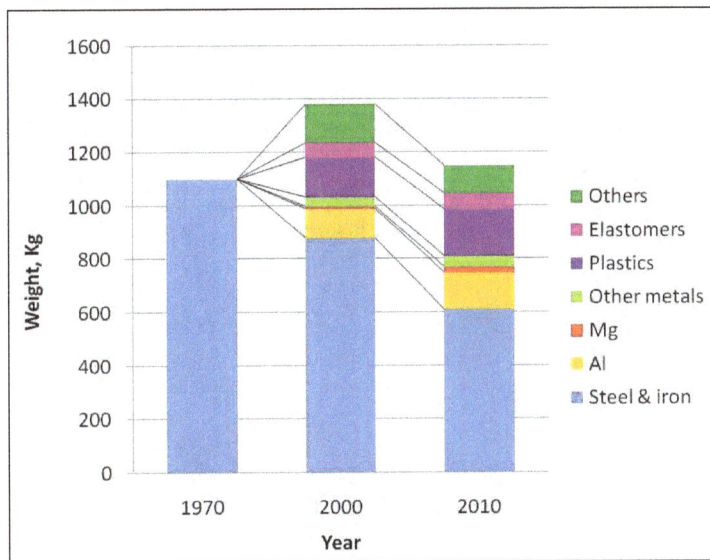

Materials used for manufacturing cars.

Ferritic Stainless Steel in Automobiles

Interstitial free ferritic stainless steels with stabilization elements like Titanium, Niobium or a combination of both are extensively being used in parts of automotive exhaust systems such as manifolds, exhaust pipes, mufflers, catalytic converters etc. In this field of application for higher efficiency, the designs are becoming more complex and performance criteria are continuously increasing. Several different grades have therefore been developed for applications at appropriate locations depending on operating condition.

Table: Typical Composition of Automotive Stainless Steels.

Grade	%C	%N	%Cr	%Mo	%Ti	%Nb
409L	0.01	0.01	11	-	~0.2	-
432	0.01	0.01	17	0.5	-	~0.2
436L	0.01	0.01	16	1	~0.3	-
439	0.01	0.01	17	-	~0.4	-
441	0.01	0.01	17.5	-	~0.15	~0.35

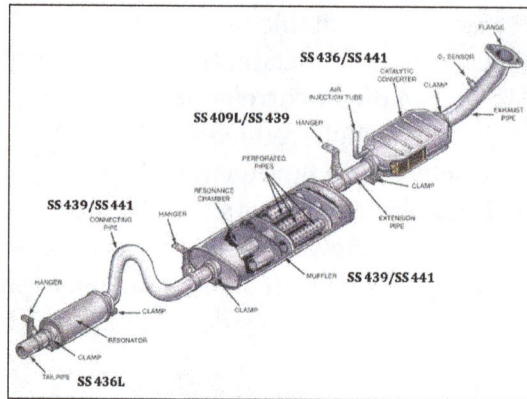

Automotive Exhaust System.

Corrosion mechanism in the hot part of the exhaust system is oxidation and in the cold part due to intermittent condensation of exhaust gases – wet and dry corrosion. Oxidation resistance of various grades is shown in figure.

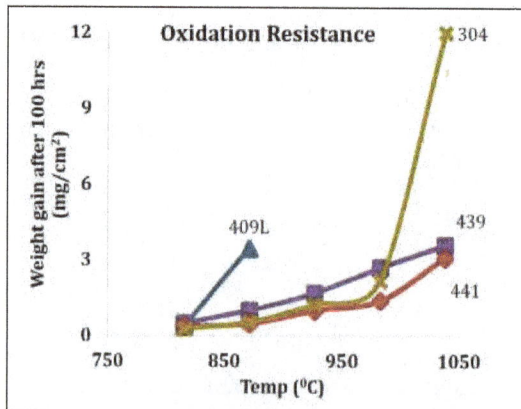

Weight gain in 100 Hrs of Oxidation in still air at elevated temperatures.

In order to simulate internal corrosion mechanism of exhaust system environment, Dip-Dry test is done. The following graph in figure allows us to compare grades and thus to improve grade selection depending upon requirements.

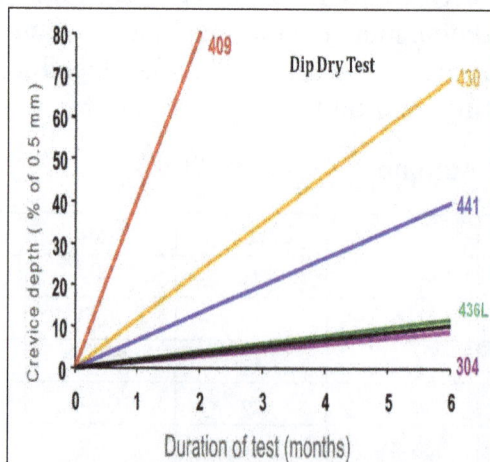

Dip Dry Test.

In components subjected to elevated temperature, creep deformation can occur. Creep- Sag test is a rough but simple method to evaluate creep behavior of a grade compared to uni-axial creep to rupture or creep to deformation tests. Creep deflection on 1.5 mm thick specimens at 850 0C, 950 0C & 1000 0C for 100 hrs are shown in figure.

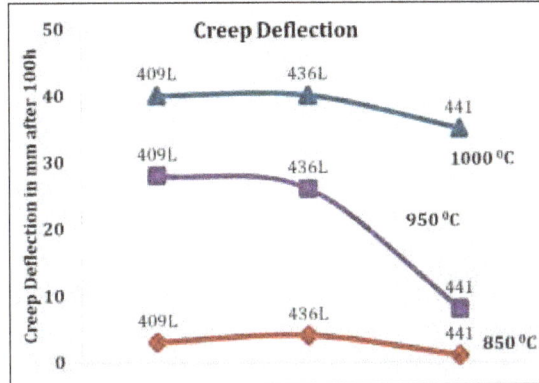

Creep Deflection at different temperatures.

Austenitic (Cr-Mn) Stainless Steel

Nickel free Chrome-Manganese Stainless steel JSLT has been successfully tested at Ashok Leyland for making of bumper that was previously being made from EDD steel (Extra deep drawing carbon steel). JSLT sheet of 1.2 mm thickness has successfully replaced 3 mm EDD steel thus providing cost and material savings.

Table: Typical Chemical Composition of JSLT.

Grade	C	Cr	Mn	Cu	N
JSLT	0.1 max	15.0-16.0	9.5-10.5	1.5-2.0	0.1-0.25
Typical composition	0.098	15.2	10.2	1.8	0.17

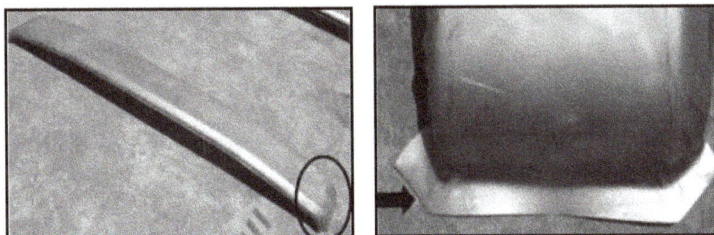

Bumper made of JSLT.

This grade has good potential for other applications as well:

• Frame, load bearing floor panels, reinforcements

• Sheet metal cabin components, body panels, A- & B-pillar, all over beam

• Fuel tanks

• Wheels, suspension arm, gear shafts, propeller shafts.

Stainless Steel in Ultralight Urban Bus

Cold worked Cr-Mn Stainless steel such as Nitronic 30 (15Cr-1.5Ni-8Mn- 0.18N) is now being used for manufacture of full size urban transit buses. This has resulted in a bus having a gross vehicle weight of 11 Tons which is less than half of a conventional transit bus.

Table: Specified Mechanical Properties for Ultra-Light bus.

	Y.S (in MPa)	% Elong.
Cold Rolled Stainless steel	800	25

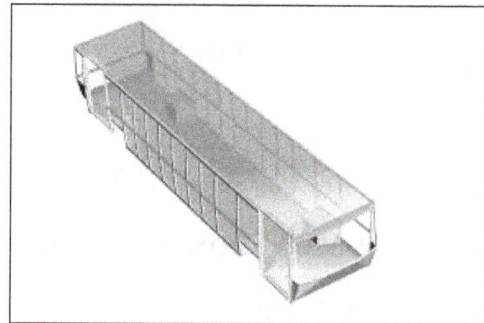

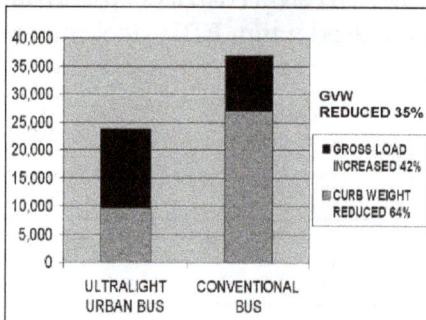

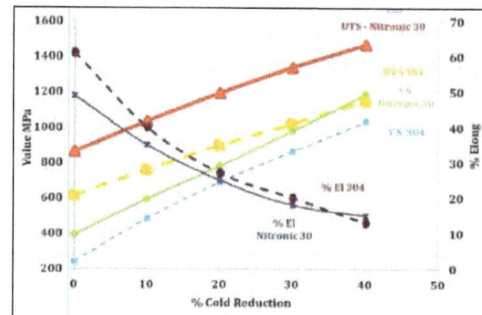

Due to the reduction in weight, 40% more passengers can be carried on the same bus. Such CrMn-N stainless steels combine very high strength with high ductility on account of transformation induced plasticity & their low electrical and thermal conductivity renders them highly suitable for spot welding.

Use of cold worked high strength Cr-Mn-N stainless steel coupled with manufacturing process such as roll forming and spot welding has led to reduction in cost of bus structure made up of such stainless steel to two-thirds of the cost of conventional steel bus.

Applications of Steel in Packaging

Steel has been a fixture in the packaging industry for over 200 years. When people think of the many uses of steel, they often think of industry. This includes oil, piping, and construction.

However, the beauty of steel is its versatility. So with its durability and portability, it's no wonder major manufacturers took to it for packaging in the 19th century.

Manufacturers and distributors in the 1800s were fascinated by steel packaging's ability to preserve things indefinitely. Many products were canned in this time period as they were transported via ship over long distances – and for long periods.

This was primarily going on in Europe, but American industry didn't take long to catch on. In 1877, Sherwin Williams patented his "ready to use paint", a revolutionary and ingenious invention at the time.

He turned to steel cans as they could be resealable, and keep the paint fresh. This trend caught on like wildfire. The first beverage can in 1935 followed in the footsteps of steel packaging.

To this day, no packaging material can compete with steel. It prevents against the weathering effects of humidity, UV-light, and gases. Furthermore, it has become a staple in pressurized aerosol products (such as cleaners, hairspray, etc.)

What's more, people fell for its appearance. Steel packaging is sleek, clean, and dynamic. It can be decorated with a paper label or directly printed on. Its glossy appearance makes it undoubtedly multi-faceted.

Many modern day manufacturers use methods like embossing and debossing, along with matte lacquers, to give steel packaging a more upscale finish.

Steel packaging's understated appearance and unobtrusive nature makes it a great medium for brands to evoke emotion and experiment with their branding.

Contemporary printing techniques only make the ability to stylize steel packaging more accessible. However, steel's longevity in the packaging industry is not merely based on aesthetics.

As individuals become more environmentally conscious, they seek out more typically "green" packaging. What many people still overlook, however, is steel's recyclability.

In Europe, over the last 10 years, steel packaging has been one of the most recycled materials. Americans also recycle 71% of steel cans. Educating more people about the ability to recycle steel cans (and how to tell them apart from aluminum) will keep its presence in the packaging industry alive.

Steel packaging manufacturers must continue to boast about their product's reusability. As a result, they will surely keep its presence alive. With its unrivaled provision of better shelf life, there's no doubt steel packaging speaks to every market – and every household in America.

Applications of Structural Steel in Energy Production

Steel is a crucial material in building a reliable energy network. Structural steel is used across various industrial sectors such as manufacturing, energy, construction, and transport. Among all these sectors, the energy sector uses the maximum amount of structural steel. Structural steel plays a vital role in the oil and natural gas industry. Structural steel fabricators at Northern Weldarc, spend a considerable amount of time in steel construction for the energy sector. There are various steel products used in the areas of energy production, transportation, and in the energy consumption facilities.

Here we are going to focus on some of the key areas where structural steel is heavily used in the energy sector.

Development

Drill rigs are massive structures used for drilling holes in the earth's subsurface. They are used to drill oil wells and natural gas extraction wells. There are various steel products used in drilling rigs such as the riser pipe, drill pipe, casing, and high tensile strength steel plates. Most of these products are fabricated by structural steel fabricators and then assembled to form a drilling rig.

Extraction

The process of extracting and producing oil and natural gas is complex. Offshore production systems are used to carry out this function. These comprise of various marine structures which are made using structural steel products and they play a role in recovering oil and natural gas. Steel tubes, steel platform, line pipe, and casing are used in equipment which extracts oil and gas.

Transportation

Transportation of the recovered oil and gas is an essential function in most rigs, especially when the consuming region is far from the production site. Oil tankers, LNG carriers, and pipelines connecting the production and consumption site are the means of transport in the energy sector. These marine carriers are built of line pipe plates and corrosion resistant materials for shipbuilding.

Storage

The main equipment used in the storage of oil and natural gas are oil tanks and gas holders. Steel plates of high tensile strength are used in this equipment.

Refining Plant

In thermal power plants, large quantities of steel pipes and tubes are in various piping applications. Some of the equipment used in refining are planted piping, heating furnace piping, and pressure vessels. The steel products used in this equipment are steel tubes and steel plates.

In Renewable Energy Production

Renewable energy technologies also rely on steel for the manufacturing:

- Solar - Steel is used in photovoltaic systems, solar thermal power plants and heat exchangers.

- Hydroelectric - Hydroelectric dams depend on steel for strong reinforcement in order to prevent dam failures.

- Wind - Steel is the primary material used in onshore and offshore wind turbines, used in nearly every part of the turbine.

- Wave and Tidal - In order to withstand the rough marine environment, strong steel components are required in wave and tidal power systems.

Wind turbine.

Applications of Stainless Steel

Stainless steel is a versatile material From the smallest zipper to the largest skyscraper, stainless steel is an essential part of modern life.

Stainless steel's strength, resistance to corrosion and low maintenance make it the ideal material for a wide range of applications. It also has a long life cycle and is 100% recyclable.

There are over 150 grades of stainless steel, of which 15 are commonly used in everyday applications. Stainless steel is made in various forms including plates, bars, sheets and tubing for use in industrial and domestic settings.

A wide spectrum of industries rely on stainless steel including construction, automotive and more. For many applications it's simply the most effective solution.

Uses of Stainless Steel

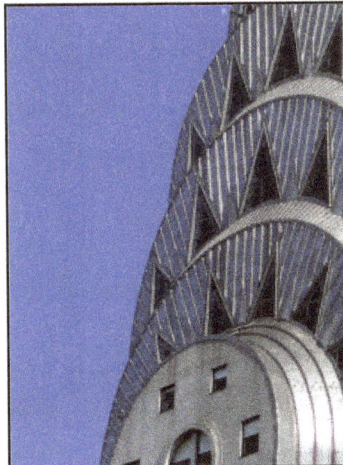

Stainless Steel accents on the Chrysler Building.

Architecture and Construction

Stainless steel first came to prominence in construction during the art-deco period. Famously, the upper portion of the Chrysler Building was constructed from stainless steel.

Due to its strength, flexibility and resistance to corrosion, stainless steel is now commonly used in modern construction. It is used in the exterior cladding for large high impact buildings and can be seen in the interiors too in the form of handrails, counter tops, backsplashes and more.

Stainless steel is easily welded, has an attractive finish and is low maintenance. Because of this, it is featured prominently in high-profile modern architecture including the Eurostar Terminal in London's Waterloo Station, the Helix Bridge in Singapore, and the One World Trade Center in New York.

The trend towards sustainable building also favors stainless steel, which is often comprised of 90% recycled metal. Stainless steel in a polished or grain finish can help bring natural light into the building, thus reducing energy consumption.

Automotive and Transportation

Stainless steel was first used in the automotive industry in the 1930s by the Ford Motor Company to make various concept cars.

Today, the use of stainless steel in the automotive industry is increasing. It is traditionally used in car exhaust systems, trim and grills, but new emission reduction standards and environmental concerns are driving manufacturers to favor stainless steel in structural components too.

Stainless steel is used in all forms of transportation including ship containers, road tankers and refuse vehicles. It is excellent for the transportation of chemicals, liquids and food products. Its high strength allows for thinner containers, saving fuel costs, while its corrosion resistance reduces cleaning and maintenance costs.

Medical

Stainless steel is ideal for hygienic environments as it's easily sterilized and resistant to corrosion. It is used in the construction of surgical and dental instruments, kidney dishes and operating tables, as well as other medical equipment such as cannulas, steam sterilizers and MRI scanners.

Surgical implants use stainless steel, as well as replacement joints such as artificial hips. Stainless steel pins and plates are used to fix broken bones in place.

Energy and Heavy Industries

The chemical, oil and gas industries operate in demanding environments involving high heat and highly toxic substances. Special grades of stainless steel have been developed for use in these industries which feature enhanced resistance to corrosion over a wider range of temperatures. High-grade stainless steel is vital in the construction of storage tanks, valves, pipes, and other components.

Super duplex steel is often used due to its high strength. It can be produced in large sheets which minimizes welding and maximizes structural integrity. Its higher strength also reduces the need for extra structural support and foundations, reducing construction costs.

Stainless steel is essential for off-shore oil rigs. Crude oil is extremely corrosive and modern rigs are constructed from high alloyed steel which is tough and lightweight.

Renewable energy technologies including solar, geothermal, hydro and wind power also use stainless steel components as it is able to withstand the rigors of highly corrosive sea water environments.

Food and Catering

Stainless steel is used in the kitchen accessories, cutlery and cookware. Less ductile grades of steel are used to make knife blades with sharp edges. More ductile grades of steel are used for items that have to be molded into shape such as cookers, grills, sinks and saucepans. Stainless steel is also used as a finish for refrigerators, freezers, countertops and dishwashers.

Stainless steel is ideal for food production and storage as it does not affect the flavor of the food. Stainless steel's corrosion resistance is important as some foods, like orange juice, can be acidic. Also stainless steel is easily cleaned which helps keep undesirable germs at bay.

Stainless steel is also important in ice cream production as it allows strong anti-bacteriological cleaning products to be used.

References

- Use-of-steel-in-building-design-construction: ukessays.com, Retrieved 3 April, 2019

- 5-uses-steel-construction: ezimetal.com.au, Retrieved 20 March, 2019

- Transport, steel-markets: worldsteel.org, Retrieved 27 June, 2019

- Steel-packaging-industry: fedsteel.com, Retrieved 7 January, 2019

- Structural-steel-energy-sector: northern-weldarc.com, Retrieved 11 May, 2019

- Most-common-uses-of-stainless-steel: metalsupermarkets.com, Retrieved 19 February, 2019

Permissions

All chapters in this book are published with permission under the Creative Commons Attribution Share Alike License or equivalent. Every chapter published in this book has been scrutinized by our experts. Their significance has been extensively debated. The topics covered herein carry significant information for a comprehensive understanding. They may even be implemented as practical applications or may be referred to as a beginning point for further studies.

We would like to thank the editorial team for lending their expertise to make the book truly unique. They have played a crucial role in the development of this book. Without their invaluable contributions this book wouldn't have been possible. They have made vital efforts to compile up to date information on the varied aspects of this subject to make this book a valuable addition to the collection of many professionals and students.

This book was conceptualized with the vision of imparting up-to-date and integrated information in this field. To ensure the same, a matchless editorial board was set up. Every individual on the board went through rigorous rounds of assessment to prove their worth. After which they invested a large part of their time researching and compiling the most relevant data for our readers.

The editorial board has been involved in producing this book since its inception. They have spent rigorous hours researching and exploring the diverse topics which have resulted in the successful publishing of this book. They have passed on their knowledge of decades through this book. To expedite this challenging task, the publisher supported the team at every step. A small team of assistant editors was also appointed to further simplify the editing procedure and attain best results for the readers.

Apart from the editorial board, the designing team has also invested a significant amount of their time in understanding the subject and creating the most relevant covers. They scrutinized every image to scout for the most suitable representation of the subject and create an appropriate cover for the book.

The publishing team has been an ardent support to the editorial, designing and production team. Their endless efforts to recruit the best for this project, has resulted in the accomplishment of this book. They are a veteran in the field of academics and their pool of knowledge is as vast as their experience in printing. Their expertise and guidance has proved useful at every step. Their uncompromising quality standards have made this book an exceptional effort. Their encouragement from time to time has been an inspiration for everyone.

The publisher and the editorial board hope that this book will prove to be a valuable piece of knowledge for students, practitioners and scholars across the globe.

Index

www.ingramcontent.com/pod-product-compliance
Lightning Source LLC
Chambersburg PA
CBHW082057190326
41458CB00010B/3517